HURRICANE
RESISTANT
CONSTRUCTION
MANUAL

Developed by
Southern Building Code Congress International, Inc.
For The
National Oceanic and Atmospheric Administration

Fredonia Books
Amsterdam, The Netherlands

Hurricane Resistant Construction Manual

by
Southern Building Code Congress International, Inc.

for
The National Oceanic and Atmospheric
Administration

ISBN: 1-4101-0883-X

Reprinted from the 1985 edition

Fredonia Books
Amsterdam, The Netherlands
http://www.fredoniabooks.com

DISCLAIMER

Reference to commercial products in this
manual do not reflect an endorsement by the
National Oceanic and Atmospheric Administration

PREFACE

The properties along the American Coast are one of the most sought after properties in the United States. People are swarming to buy or build on beach property on the Atlantic Gulf and Pacific Coast. The increase in population and building has created many problems. Beaches are getting smaller and building density is increasing. This increasing density increases the changes for large developed areas to be in the path of a hurricane and the larger the developed area, the greater the potential for property loss and deaths.

Both coastal development will continue and hurricanes be spawned and make landfall. If we are to minimize property loss along our coast, we must construct buildings that will resist the water and wind forces in hurricanes.

The purpose of this manual is to provide design and construction, details and technique for improving coastal construction.

TABLE OF CONTENTS

CHAPTER I

ELEMENTS OF HURRICANES

I. **INTRODUCTION**

The most important step in the design and construction of a building is the determination of the types and magnitudes of natural hazards that will impact a building during its lifetime. On much of the Gulf of Mexico and Atlantic coastlines, hurricanes, with their associated high winds greater than 74 mph, storm surges, and large waves are the hazard of greatest concern. Earthquake hazards must be accounted for in a limited number of coastal areas, such as in South Carolina. North of South Carolina, winter storms, or nor'easters, can be equally or more damaging than hurricanes because of their greater frequency, longer duration, and their high erosive impacts on the coast. Although earthquakes and nor easters are important to consider, this manual is limited to and specifically addresses the problems of hurricane-resistant design and construction for residences and small commercial structures.

Once the hazard is identified, it is then necessary to establish design criteria for the building. To establish these criteria some statistical analyses is necessary. In the case of hurricanes, many of the analysis have been made, including recurrence interval, wind velocities, storm surge, and wave heights.

II. **RECURRENCE INTERVAL**

Building codes require a minimum design condition using a wind speed based on either a 50 year or 100 year mean recurrence interval (MRI). The recurrence interval is based on records maintained over a long period of time. For example, if records were kept between 1885 and 1985 and a wind velocity of 100 mph occurred in the years 1934 and 1936, the recurrence interval would be 50 years for a 100 mph wind. A 50 year MRI means that each year there is one chance in 50 that the mapped wind speed will occur. While one chance in 50 seems small at first glance, the probability of such a storm over the lifetime of a residence or other building is cumulative. Statistically, the probability of a 50 year storm hitting in any 50 year time period is about two chances in three. Table 1-1 indicates the probability or percent chance that events of different sizes and rarity will occur at least once in a 10- to 70-year period. Note that even for the extremely rare, catastrophic event such as 200-year storm, the probabilities are nearly 3 in 10 (30%) that the event would occur in a 70-year building lifetime.

TABLE 1-1
OCCURENCE PROBABILITY

Event	Probability of Events Occuring in a 10 - 70 Year Period Percent Chance of Occurring at Least Once In				
Annual Probability	10 yrs	20 yrs	30 yrs	50 yrs	70 yrs
10-year	65%	88%	96%	99%	99.9%
25-year	34	56	71	87	94
50-year	18	33	45	64	76
100-year	10	18	26	39	51
150-year	06	13	18	29	37
200-year	05	10	14	22	30

Generally a residence lifetime is expected to be approximately 70 years. If a residence is to be designed for a 100 year recurrence interval, the chance of the residence experiencing the storm during its expected lifetime is 51 percent, and the same residence for a 50 year recurrence is 76 percent.

A 100 year or 50 year recurrence may sound like a severe design condition for a building. However, when one considers that a building has a 50% or 75% chance of being required to resist these loads, it is a very reasonable minimum standard. In fact, a building design for loads in excess of these minimum values may be a wise investment.

III. WINDS

Wind velocities have been recorded for more than a century at weather stations throughout the United States. Most of these stations are located several miles inland from the coast. Ground surface friction is much higher than water surface friction and as a result, the wind velocities at the inland stations may be considerably smaller than those at shoreline sites. This difference can be as much as 30 percent for sites 5 to 7 miles inland. Most wind velocities shown in codes have accounted for this difference. Maps indicating the wind velocities for 100 and 50 year recurrences are shown in Figures 1-1 and 1-2, respectively. A comparison of the two maps for Key West, Florida indicates the 100 year interval map Figure 1-1 shows a design wind velocity of 130 mph, while the 50 year map shows a design wind velocity of 110 mph. It should be noted that mathematical comparisons between the two maps cannot be made as the maps are based on different data.

Generally local governments adopt one of the model codes for regulating construction. Each of the model codes specify one or both of the Basic Wind Speed Maps shown in this section. When both maps are specified, the 50 year recurrence interval map is specified for building heights of 60 feet or less. In areas where local data indicates the design wind velocities are different from those indicated on the two maps, the local government normally adopts one or two specific design wind velocities for the area.

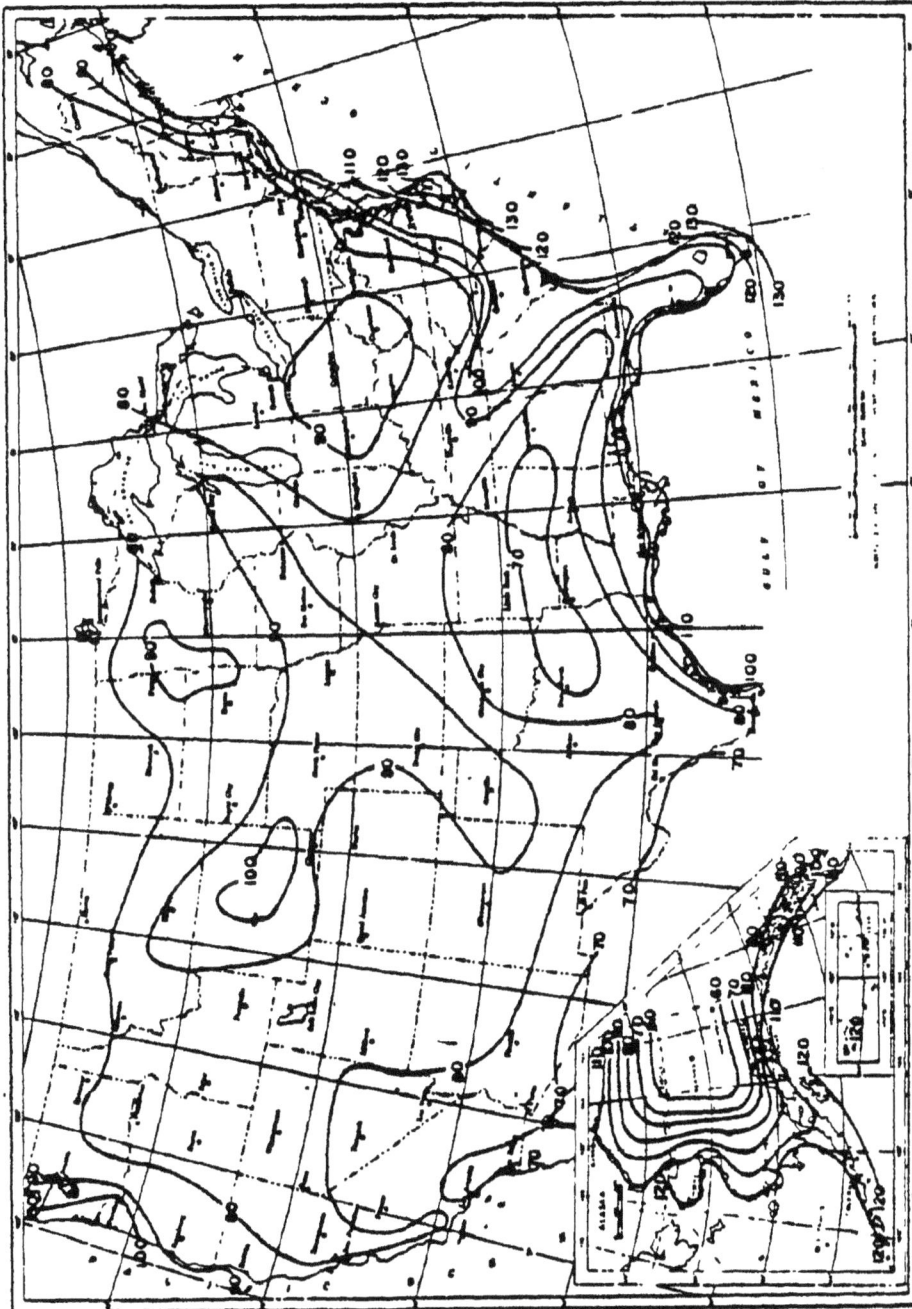

Annual Extreme Fastest-Mile Speed 30 ft Above Ground, 100 Year Mean
Recurrence Interval

NOTE: The Virgin Islands and Puerto Rico shall use a basic wind speed
of 110 mph.

(SOURCE - 1985 EDITION STANDARD BUILDING CODE)

FIGURE 1-2
BASIC WIND SPEED IN MILES PER HOUR –
50 YEAR MEAN RECURRENCE INTERVAL

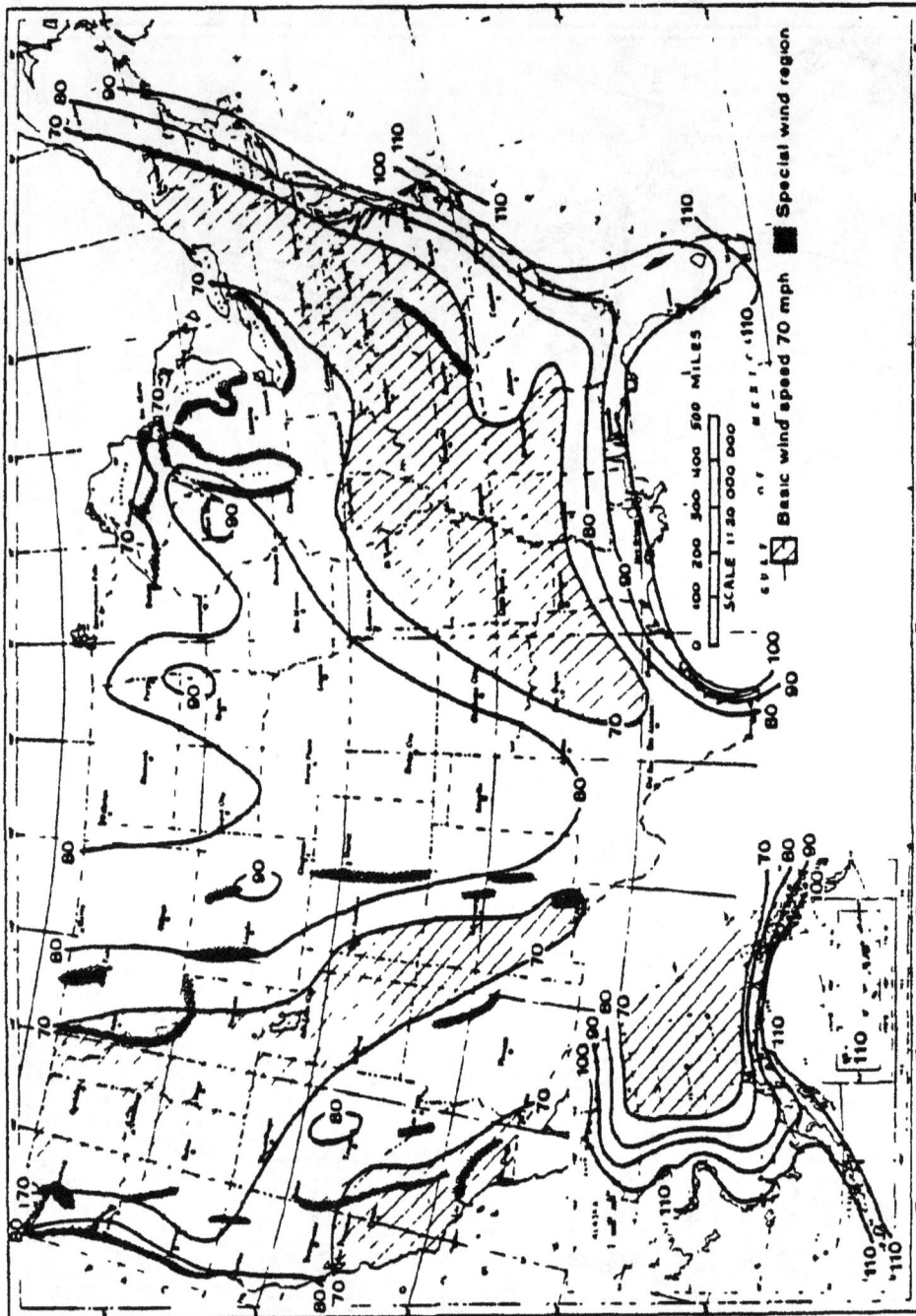

1. Linear interpolation between wind speed contours is acceptable.
2. Caution in the use of wind speed contours in mountainous regions of Alaska is advised.

(SOURCE – 1985 EDITION STANDARD BUILDING CODE)

The fastest mile wind speed is defined as the highest speed at which a mile of wind passes a measuring point 30 feet above the ground. This wind speed differs from the wind speed reported by the National Hurricane Center, and the news medium. The wind speed reported by the center and news media is peak gust associated with a one or two second averaging time. This results in reported wind velocities being considerably higher than equivalent design wind velocities. The paper "Wind Speed-Damage Correlation in Hurricane Frederic" compares calculated equivalent fastest mile wind velocity to measure peak gust velocities in Hurricane Frederic. This comparison in Table 1-2 illustrates a significant difference in the two types of wind values.

TABLE 1-2
COMPARISON OF REPORTED AND DESIGN WIND VELOCITIES

	Location	Height of anemometer in feet	Reported Measured[a] Peak gust, in miles per hour	Design Calculated Fastest Mile[b] Speed, miles per hour
1	Gulfport Civil Defense	89	95	79
2	Ingalls Ship-	33	127 (est)	107
3	Dauphin Island Bridge	60	145	106
4	Mobile Airport	22	97	87
5	Mobile Civil Defense	75	109	90
6	Coast Guard Cutter White Pine	49	117	95
7	Pensacola Naval Air Station	75	95	74
8	Pensacola Regional Airport	22	78	73
9	Eglin Air Force Base	15	49	50

a. Peak gusts are associates with one to two second averaging time (based on anemometer and recording system characteristics).
b. Converted to fastest mile speed at 33 ft. (10 m) above ground in flat, open terrain.

(Source - ASCE Paper "Wind Speed - Damage Correlation in Hurricane Frederic by K. C. Mehta, J. E., Minor and T. A. Reinhold)

This lack of equating reported wind velocities to the fastest mile wind velocities has resulted in numerous accusations that local, state and model code wind requirements are inadequate. For example, the peak gust wind velocity at Dauphin Island during Hurricane Frederic was reported at 145 mph (Table 1-2). When one compares the reported velocity and the design wind velocity of 110 mph (Figure 1-1), one would believe that the hurricane winds exceed the design requirements. If, however, the peak gusts are equated to a fastest mile wind, a velocity of 106 mph (Table 1-2) is obtained and we find the hurricane winds did not exceed the design wind requirements. This example clearly points out the need for local and state governments to equate maximum recorded peak gust winds to fastest mile winds prior to changing wind design requirements in codes.

In the northern hemisphere, hurricane winds spiral counterclockwise around an eye, a low-pressure area in which wind speeds are only 10 to 20 mph. The winds are strongest just outside the eye on the upper right side or quadrant, because the forward movement of the storm is added to wind speeds of the storm. The wind velocities decrease as the distance from the eye increases. As the eye approaches a site, the winds increase gradually to a peak before the eye passes the site, drops to a calm as the eye passes over the site, and increases to another but lower peak on the back side of the eye. The winds on the back side blow in the opposite direction to those on the forward side as a result of the circular wind pattern in the hurricane. Inasmuch as buildings may be exposed to winds from several directions as the storm approaches and passes, buildings must be designed to resist winds from all directions.

IV. STORM SURGE

As a distant storm approaches the coast, oceanic swells generated by the storm cause a slow rise in water level, known as the "forerunner"; the rise can be as much as 3 or 4 feet over several hundred miles of coast. Storm surge is a rapid rise in water level above normal water level, caused by hurricane winds and decreasing barometric pressure. Storm surge can be pictured as a cone of water, a bulge caused by the low barometric pressure of the storm and the stress or friction of storm winds on the surface of the ocean. As the hurricane approaches land and shallow water, the water level rises rapidly in a mound above normal levels expected for any given stage of the astronomical tide. Storm surge rises can range from 4 - 5 feet in a Category 1 storm (74 - 95 mph winds) to greater than 18 feet in a catastrophic Category 5 storm (greater than 155 mph winds). Maximum storm surge usually occurs 10 to 20 miles to the right of the storm track; during Hurricane Camille in 1969 maximum surge of 25 feet above mean sea level occurred about 10 to 15 miles east of the eye's landfall at Waveland, Mississippi. However, maximum surge may occur to the left of a storm if the winds pile water against an obstruction such as the landward side of a coastal barrier island.

The shape of the land and the slope of the sea bottom are major factors in the type of a surge created by a storm. Large open bays in the path or to the right of the storm will have water pushed into the entrance by the storm. As the surge reaches the head of the bay it is constricted and can only rise, considerably higher than at the entrance. Water as deep as 10 feet flowed through and destroyed the Brownwood Subdivision of Baytown, Texas at the head of Galveston Bay during Hurricane Alicia. Bays to the left of the storm's eye will have water pushed seaward, and may pile up against or overflow any coastal barrier in the way. A surge moving over a straight shoreline piles water up against the shoreline escaping slowly at each end, or, in the case of low-lying coastal barriers, flowing directly over the barrier. Surge as deep as 6 feet flowed over many parts of Ft. Morgan peninsula during Hurricane Frederic in 1979. In large lakes, water is piled against one side of the lake and when the eye passes over, the wind reverses direction and a larger surge of water is pushed back toward the other side.

The tide occurring at the time a hurricane reaches landfall is another factor when coupled with the effect of storm surge. For example, if the range between low and high tides is 8 feet and an 8-foot storm surge occurs at low tide, there would be little or no net effect from rising water levels. This phenomenon was basically why the 1984 Hurricane Diana had such minimum flooding and erosive effect at landfall in North Carolina. If the same surge occurred at high tide, the water level could be nearly twice normal height for that stage of the tide.

Based on all of these factors, plus the variable frequency and paths of hurricanes, predicting storm surges on the basis of 100 year recurrence is difficult. Nevertheless, the Federal Emergency Management Agency, has developed flood plain maps for all coastal areas. In coastal areas, the velocity (V) Zone and A Zone indicate the areas that would be flooded by storm surge. The velocity zone is defined as an area subject to flooding and high velocity waves greater than three feet high in a 100 year (one percent chance) storm. The A zone is defined as an area subject to flooding and wave action of less than 3 feet in such a storm. A typical FIRM (Flood Insurance Rate Map) is shown in Figure 1-3.

FIGURE 1-3
EXAMPLE OF TYPICAL FLOOD INSURANCE RATE MAP

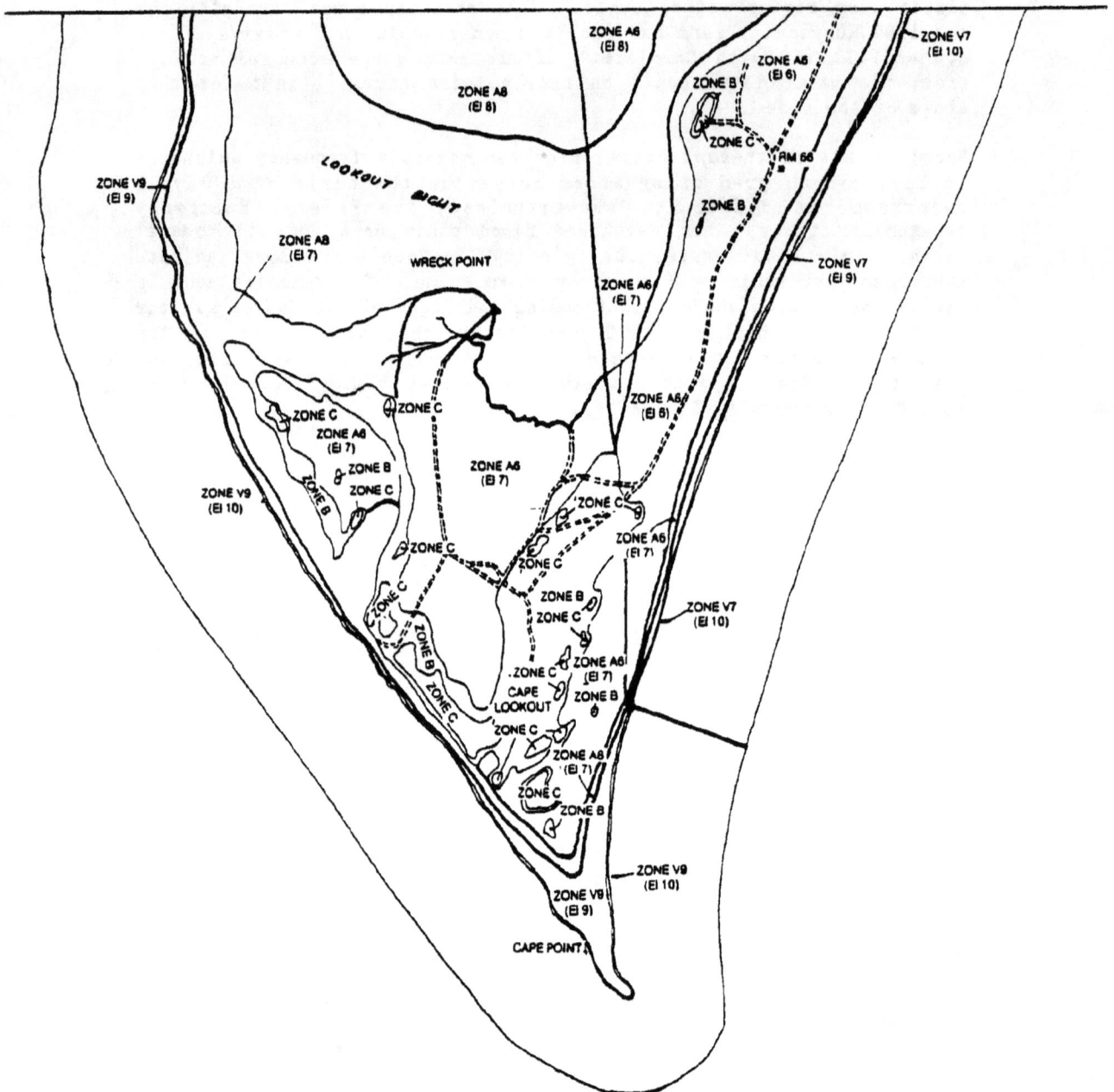

(Source - FEMA - Firm for Carteret County North Carolina,
Unincorporated Areas)

V.   WAVES

Wind generated waves are formed when the force of the wind interacts
with the surface, and builds in height and depth. The eventual height
of the wave in deep water is controlled by several factors including:
wind speed; fetch (distance over which the wind blows); duration of
the storm; and water depth. The faster the winds, the higher the

waves. The size of a storm also affects the waves. In a large area, high winds blowing over the same waves for the entire length or fetch of the storm cause the waves to continue to build in height. The time or duration of the storm in a stationary position also affects the wave heights. The longer the wind blows over the same waves, the larger they can become.

Hurricanes can build waves as high as 50 to 75 feet in the open ocean, where the water depth can support waves of that height. Waves could grow even larger in such high winds, but are limited by the relatively small area covered by the fastest winds and the continued movement of the storm, limiting the duration of waves building at any one position. On the other hand, northeast extratropical storms (nor'easters) can build equally large waves although they have considerably slower winds. Lower windspeeds are offset by very long distances or fetches of fastest winds and the duration of the storm often in nearly stationary positions for several days.

As a wave moves into shallow water, the wave's friction against the ocean bottom slows the deeper motion of the wave. The wave eventually reaches an unstable condition where the top of the wave is moving faster than the base of the wave. The top of the wave spills forward and the wave breaks. The wave is not destroyed, but reforms a smaller wave that continues toward the shore until it, too, is forced to break. Measurements taken in the lab and in the field have established that a wave is forced to break when the wave height is approximately 78 percent as high as the water is deep (See Figure 1-4). The final height of the breaking wave is generally less than 5 or 6 feet. The same process limits the wave heights on the shoreline even in a 100-year design storm. However, in design conditions, the storm surge raises the water level well above normal, increasing water depth and the size wave that can be supported.

FIGURE 1-4
MAXIMUM WAVE HEIGHT

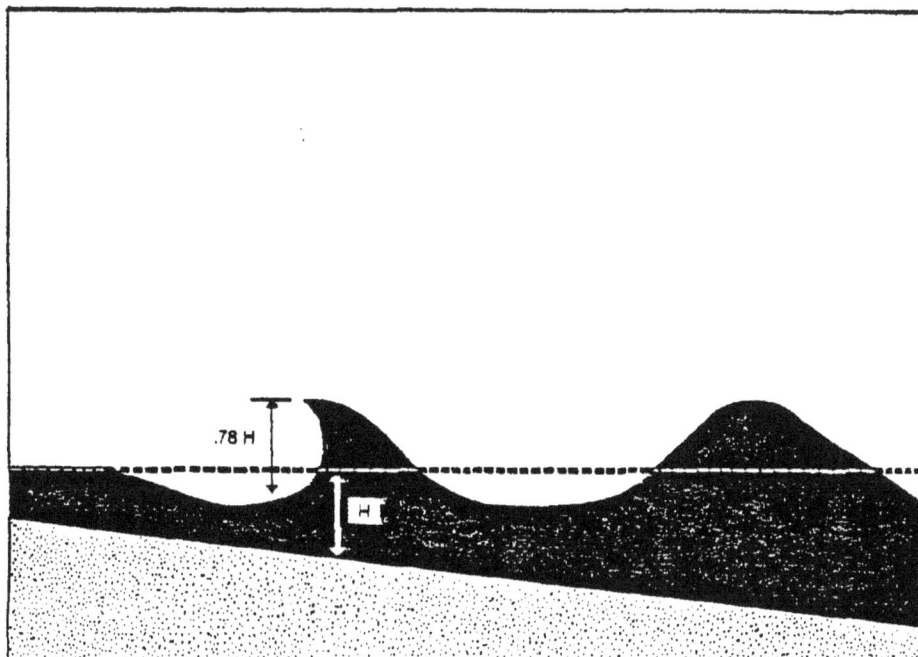

For flood insurance purposes, the Federal Emergency Management Agency identifies coastal areas subject to wave action as the V-Zone on Flood Insurance Rate Maps. The V-Zone is defined as an area of special flood hazard subject to tidal flooding with velocity. The designation is generally applied to areas where the still storm-water height (height of astronomical tide plus surge) is sufficient to support at least a 3-foot wave. The Agency has prepared a practical field manual for estimating wave heights in the Atlantic and Gulf Coast regions which is available to any community or individual wanting to know the procedures in detail. "Flood Plain Management: Ways of Estimating Wave Heights in Coastal High Hazard Area in the Atlantic and Gulf Coast Regions", TD-3, Federal Emergency Management Agency, Washington, D. C. 20472, April 1981. Designers and builders should consult with the Building Inspection Department to determine the basic flood elevation for any proposed building site.

VI. HURRICANE SCALE

Historical accounts have characterized each hurricane as the worst ever, or greater than a specific previous storm. In an effort to better classify storm, the National Weather Service uses the Saffir-Simpson Hurricane Scale (See Table 1-3) to report storms. The scale is based on 3 storm variables - wind velocity, storm surge and barometric pressure.

TABLE 1-3
SAFFIR-SIMPSON HURRICANE SCALE

Category	Wind (mph)	Storm Surge (feet)	Central Pressure (inches)	Damage
1	74-95	4-5	>28.94	minimal
2	96-110	6-8	28.50-28.91	moderate
3	111-130	9-12	27.91-28.47	extensive
4	131-155	13-18	27.17-17.88	extreme
5	>155	>18	<27.17	catastrophic

(Source - NOAA - NWS Southern Regional Technical Report 2)

All hurricanes making landfall cause damage, so no one should be misled by the hurricane scale. The wind may rapidly increase or the coastal configuration may amplify the storm surge level. The scale should be used for the purpose it was developed - the categorization of hurricanes - all of which may cause extensive damage.

VII. COASTAL CONSTRUCTION PRACTICES

Over the past ten years, numerous surveys have been made to determine the primary causes of building failures in hurricanes. These surveys have identified a number of locations where a building was totally destroyed while the adjacent buildings incurred only minor damage. Initially, it was thought the damage was caused by hurricane-generated tornados since this non-damage/damage condition has been experienced in areas hit by tornados. Further analysis of the buildings indicated that the difference occurred as the result of design and construction

differences. <u>The building that failed had inadequate pile penetration</u>
<u>while the building that incurred minor damage had adequate pile</u>
<u>penetration - or - the building that failed had inadequate</u>
<u>connections, while the slightly damaged building had adequate</u>
<u>connections.</u> These surveys proved conclusively that it is
economically feasible to design and construct one and two family
dwellings and small commercial buildings to resist the winds, surges
and waves in most hurricanes. However, recent surveys of areas hit by
Hurricane Alicia in 1983, Diana in 1984 and Elena in 1985 indicate
that many one and two family dwellings and small commercial buildings
are continuing to be designed and constructed with inadequate
attention to hurricane resistant details.

In a number of buildings surveyed, designers and builders incorporated
some hurricane resistant construction details in the design. The
primary problem appears to be that they were not aware of all of the
requirements and, like a weak link in a chain, the weak or
non-existing construction detail caused the building to fail. A good
example of this was the tying of the rafters to the top plate with
hurricane connectors and only nailing the top plate to the wall studs.
Needless to say, the roof was blown off with the top plate still
attached, and with the loss of the roof support the walls failed (See
Figure 1-5).

FIGURE 1-5
RAFTER TO PLATE CONNECTOR

Hollow masonry unit block walls of many small commercial buildings
failed for lack of vertical reinforcement bars in the walls (See
Figure 1-6).

FIGURE 1-6
UNREINFORCED MASONRY

Even those dwellings and other buildings that had little or no
structural damage often experienced internal water damages from
wind-driven rain leaking around window casements, under or around
sliding glass doors, through gable end roof vents, or other component
failures. Heavy damage to interior walls, flooring, floor covering,
wiring, insulation, and interior furnishings can result from even
minor openings of the structure or from failure of component details.

Dwellings and small commercial buildings tend to be non-engineered or
marginally engineered. They continue to be frequently constructed
using techniques not suited to the wind, surge and waves associated
with hurricanes. This manual is developed to provide these details
and techniques to minimize property losses. The manual is designed
specifically for inspectors, builders, superintendents and foremen
involved in coastal construction. In addition, homeowners will find
this manual beneficial, as it will provide them with the techniques
for coastal construction. As consumers, they will be able to demand
that these techniques be included in their coastal dwelling.

CHAPTER II

HURRICANE – BUILDING INTERACTION

I. INTRODUCTION

As a hurricane moves across the open ocean, the storm accumulates vast amounts of energy. As it makes landfall and moves inland, this vast energy is dissipated by the land areas as the storm interacts with the land features and loses its ocean energy source. The storm interaction with a building will be in the form of wind and water. When the building becomes a barrier to the free movement of the wind or water, forces will be transmitted from the wind or water to the building. A building designed and constructed to be hurricane resistant must be designed to withstand these forces. Let's look at how each of the forces interact on a building and its various components.

II. Water

Water will interact with a building in at least three ways — surge, waves and scour. Surge will interact with a building in a way similar to that of a river flood. It may cause the building to float off its foundation (Figure 2-1), and may cause floating debris to collide against the building. If the building is left intact, surge water entering the building may destroy or damage the contents, walls, flooring, insulation, wiring, and so on. Surge forces on the building are primarily the momentum of water striking the building and the drag force in the direction of the water flow (Figure 2-2), or flotation which tends to lift the pilings or produce upward forces on the first floor of a building; sometimes both.

FIGURE 2-1
BUILDING FLOATS OFF FOUNDATION

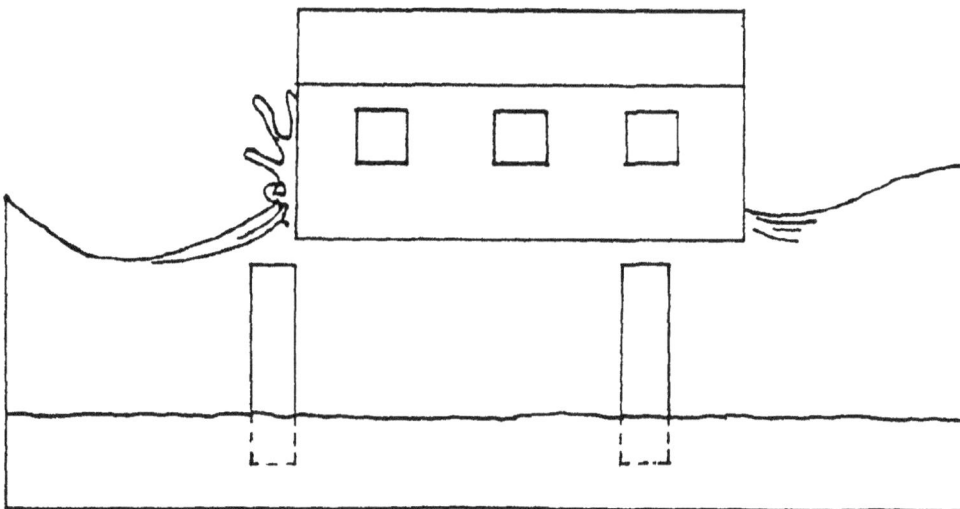

FIGURE 2-2
WATER FORCES ON BUILDING

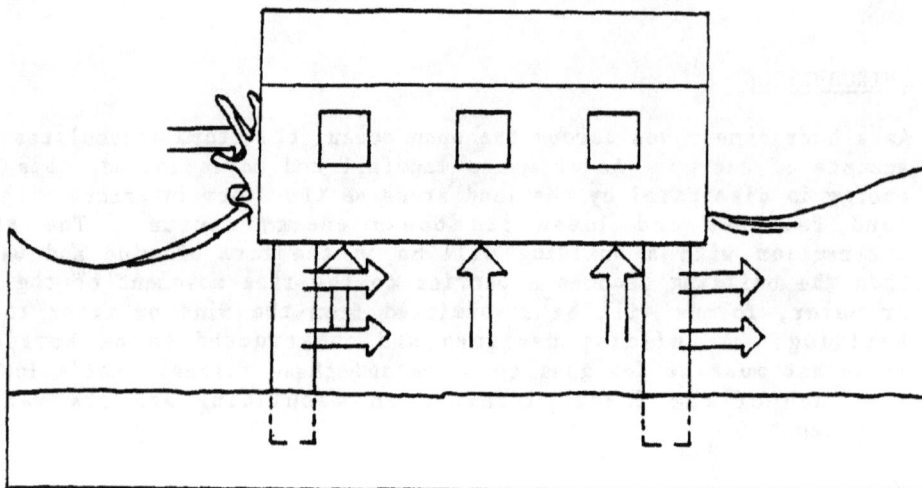

Hurricane waves increase the effective height of the storm surge as much as 55 percent higher than the storm surge level at the coast. Although surge typically has forward motion, the force of the wave is much more destructive because the velocity of the wave may be several times faster than the surge. For example, a 5-1/2 foot wave moving at a velocity of 10 mph that is stopped by a structure may produce a force in excess of 1000 psf. This force will dislodge small buildings from their foundations, boats and barges from their moorings. Objects then may be thrown against a building or other structure as debris, significantly increasing the forces exerted by the wave. Because of these large potential forces, most codes and standards require all structures to be elevated above the wave crest elevation. Elevation above the wave crest level eliminates all lateral water loads against the building itself leaving the loads to pilings for design consideration. The pilings must be designed to resist the impact forces of the waves and the drag forces of the water in the direction of the velocity of the water (Figure 2-3).

FIGURE 2-3
IMPACT AND DRAG FORCES ON PILES

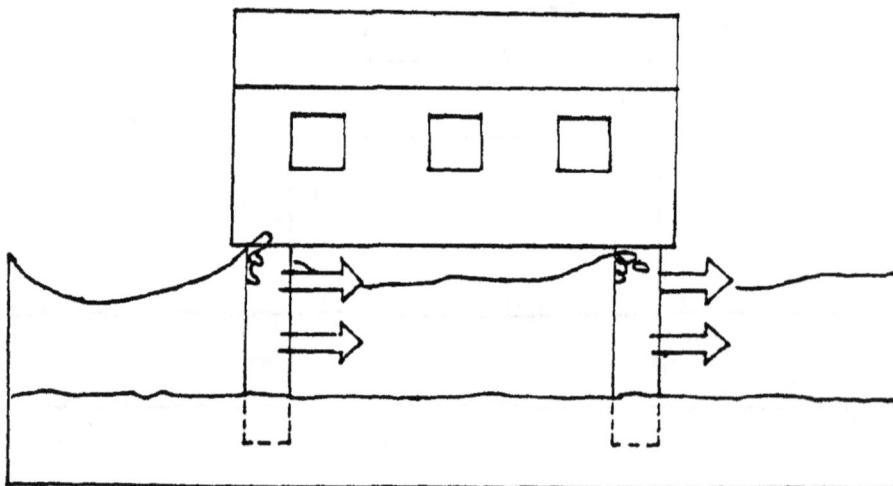

Scour is the erosion of sand and soils caused by wave action, and may penetrate landward tens to hundreds of yards in the course of a storm. On low-lying, sandy coastal barriers, scour depths of 4 to 6 feet are common, sometimes leaving well-delineated scarps, sometimes causing sand buildups from overwash several hundred yards landward. Houses or buildings in the scour zone (typically in the first row or tier from the shoreline) often have several feet of supporting soils removed from around their pilings, thus increasing the lateral forces acting on the pilings. Grade platforms in the scour zone are typically undercut, often collapsing because of inadequate reinforcement and the uplift forces of the waves under them.

While scour will generally not erode below sea level when it confronts no steep or vertical obstructions, it may undermine protective seawalls or bulkheads, eroding well below sea level. Waves striking the Holiday Inn at Gulf Shores, Alabama during Hurricane Frederic scoured a hole about 20 feet deep on the vertical face of the Inn's Gulf-front.

The need to consider scour becomes evident when one sees the consequences of neglecting to consider it. Almost all of the houses that collapsed on the Gulf shoreline on the west end of Galveston during Hurricane Alicia collapsed because of scour and inadequate piling penetration into supporting soils. Piling penetration ranged from 3 ft. 9 in. to 6 ft. below the pre-storm ground surface in areas that experienced 6 feet of scour. The buildings had to collapse; they had no ability to resist the uplift and horizontal forces of the storm once the supporting soil was scoured away.

III. Wind

Wind acting on a building produces different forces depending on the location of the various building components relative to the direction of the wind.

A. Walls - Wind acting on a rectangular structure induces inward acting pressures on only one wall, the wall facing against the wind. This wall is called the windward wall (Figure 2-4) with the opposite wall being called the leeward wall, and the two other walls being called side walls. Wind blowing against the windward wall (the wall facing the wind) will induce inward acting pressures on that wall (Figure 2-5), and outward acting pressures on the leeward wall (Figure 2-6) opposite to the windward wall) and side walls (Figure 2-7). Wind flowing around the corners of the walls is changing flow direction, and the resulting turbulence induces high localized outward pressures (Figure 2-8).

FIGURE 2-4
WALL IDENTIFICATION

PLAN

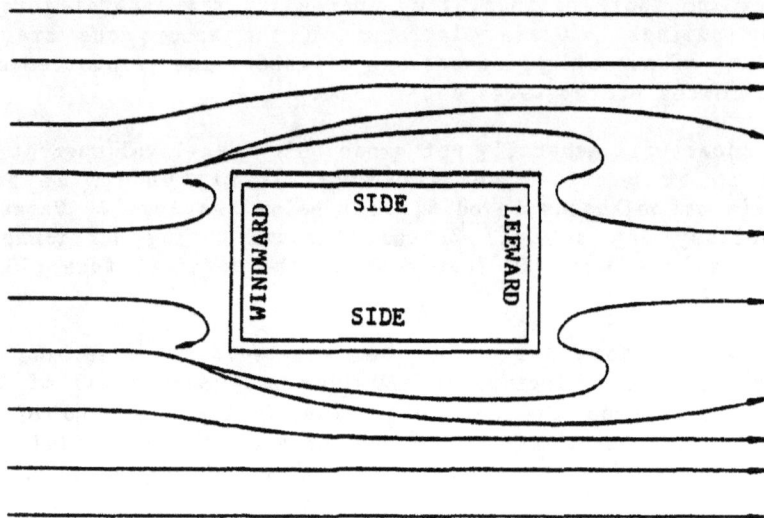

FIGURE 2-5
WINDWARD WALL LOADING

SECTION

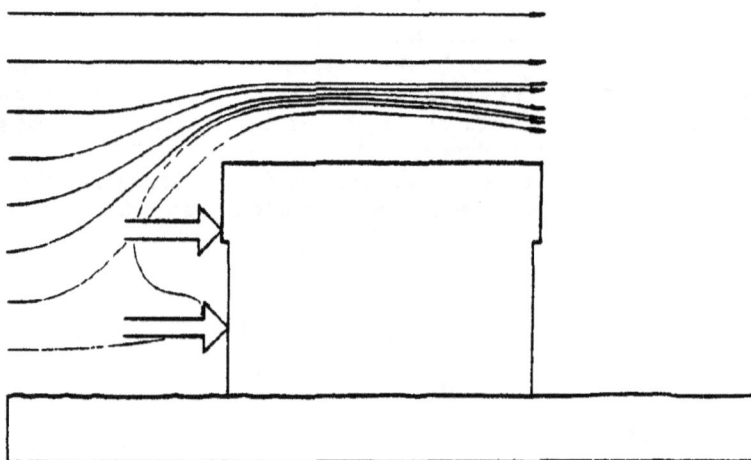

FIGURE 2-6
LEEWARD WALL LOADING

SECTION

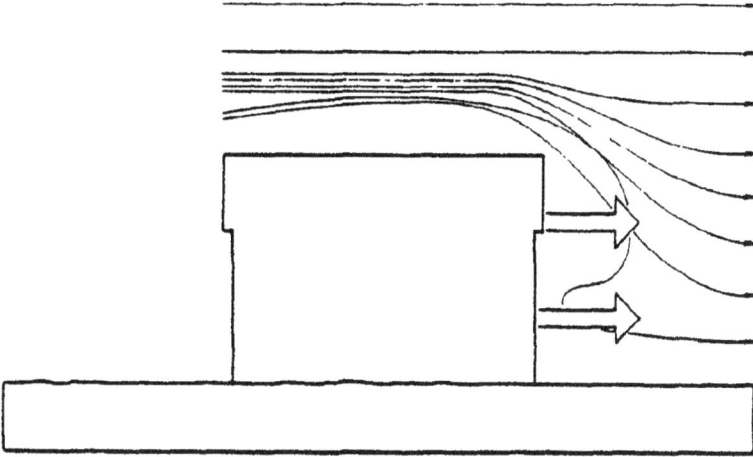

FIGURE 2-7
SIDE WALL LOADING

PLAN

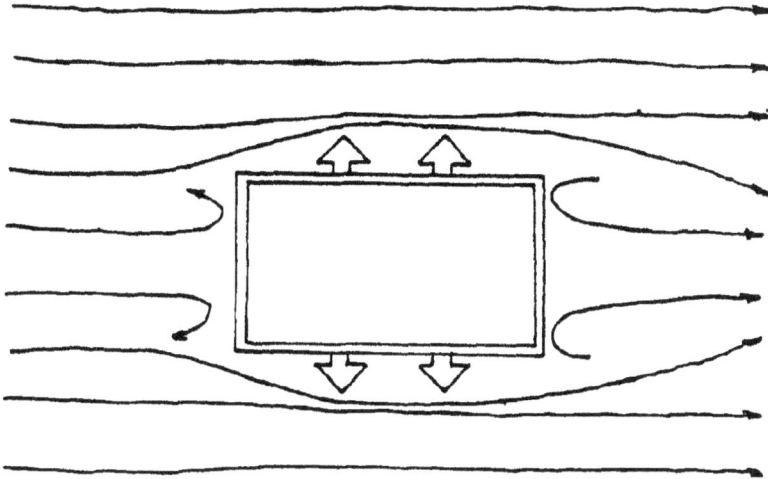

FIGURE 2-8
CORNER LOADING

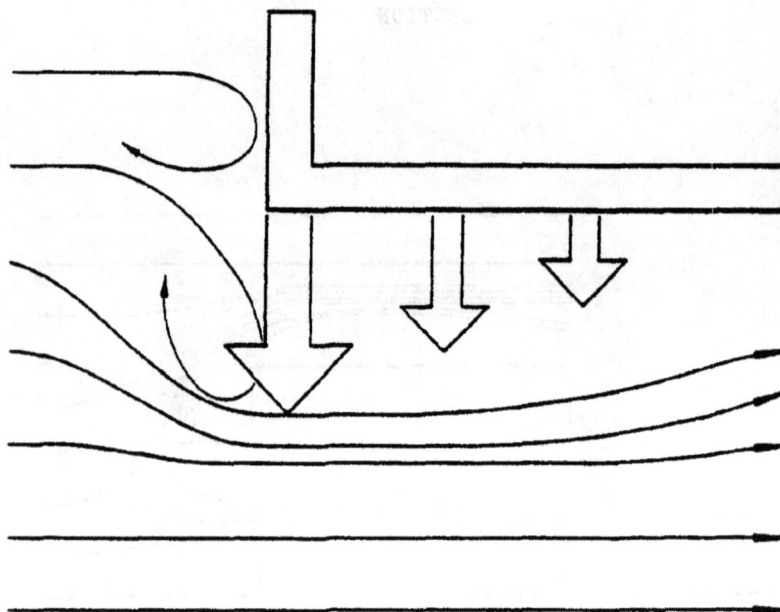

B. <u>Roof</u> - Wind acting on a roof may induce outward or a combination of inward and outward pressure depending on the wind direction and the slope of the roof. Winds interacting with a flat roof and roofs with slopes of 40° or less will have an outward pressure over the entire roof (Figures 2-9 and 2-10). A wind blowing in a direction normal (perpendicular) to the ridge with the roof slope greater than 40° will have an inward pressure on the windward side of the roof and an outward pressure on the leeward side (Figure 2-11). A wind blowing parallel to the ridge will result in outward pressures for the roof surfaces. Winds flowing around eaves ridges, and overhangs induce high uplift forces on these roof components (Figures 2-12 and 2-13).

FIGURE 2-9
FLAT ROOF LANDING

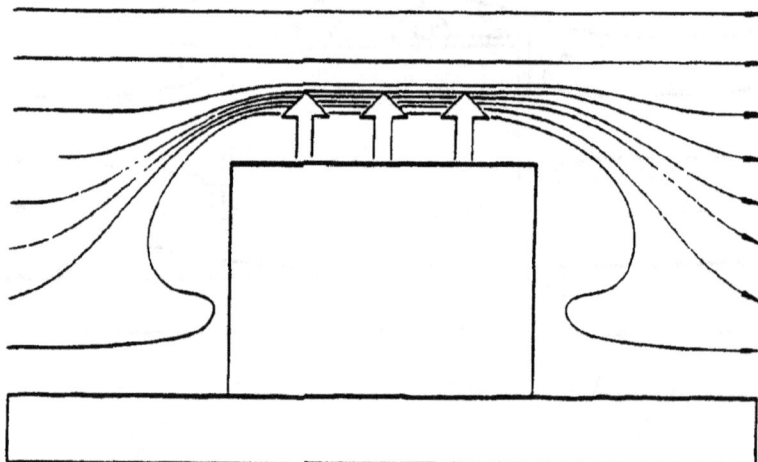

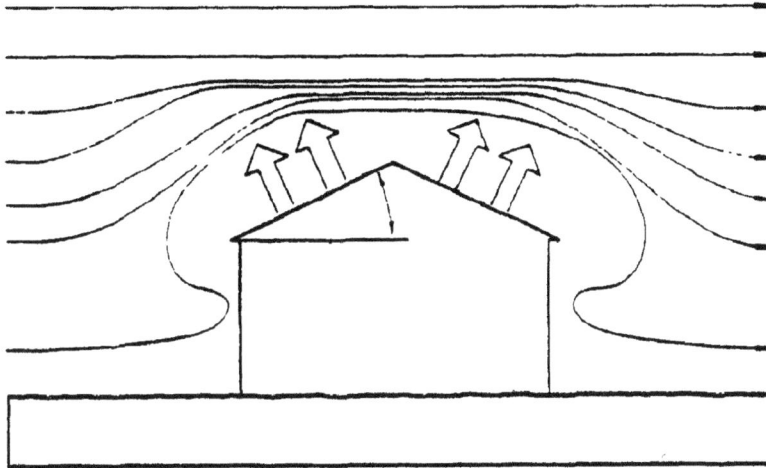

FIGURE 2-10
(Slope 40° or Less)
LOW SLOPED ROOF LANDING

SECTION

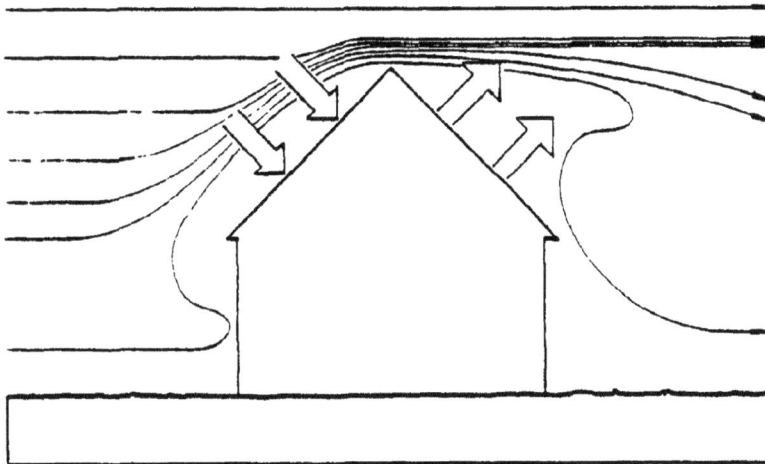

FIGURE 2-11
HIGH SLOPES ROOF LOADING

SECTION

FIGURE 2-12
OVERHANG AND RIDGE LOADING

SECTION

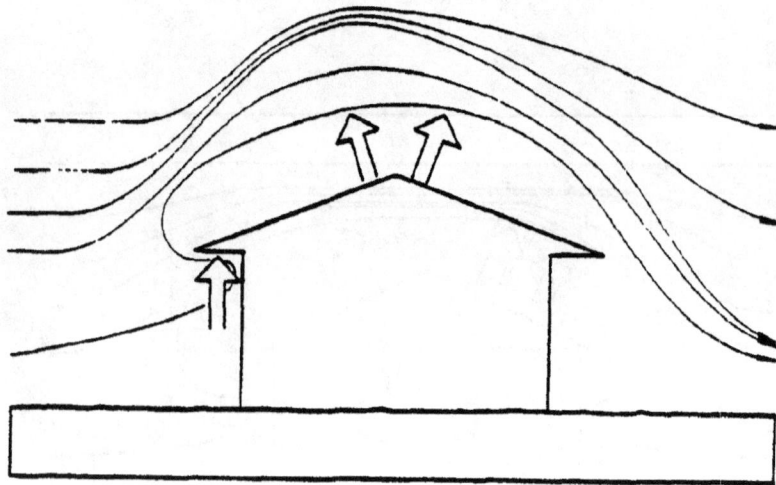

FIGURE 2-13
CORNER OVERHANG LOADING

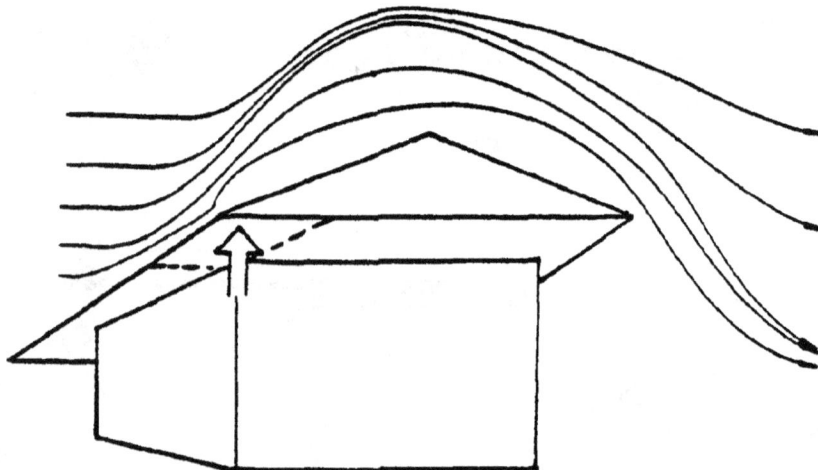

C. **Internal Pressures** - If wind gains entrance to a building through the failure of windows or doors, pressure changes are created inside the building. The induced pressures inside the building depend on the location of the failed opening with respect to wind direction. If the opening fails in the windward wall the pressure inside the building increases (Figure 2-14), and if the failure is in the side or leeward walls the pressure inside the building decreases (Figure 2-15). Figures 2-16 and 2-17 show the internal pressures on the roof as the result of a windward and leeward wall failure respectively. Even without opening failures, there are significant internal pressure changes through normal cracks and crevices in the building.

FIGURE 2-14
INTERNAL WALL PRESSURES AS THE RESULT OF WINDWARD OPENING FAILURE

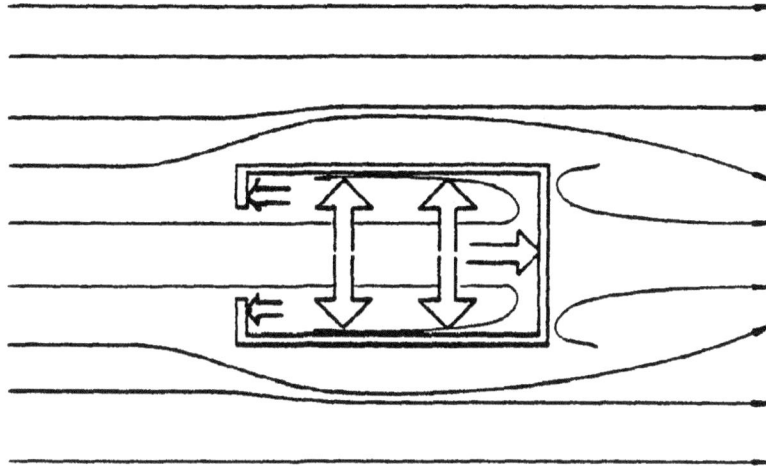

FIGURE 2-15
INTERNAL WALL PRESSURES AS THE RESULT OF LEEWARD OPENING FAILURE

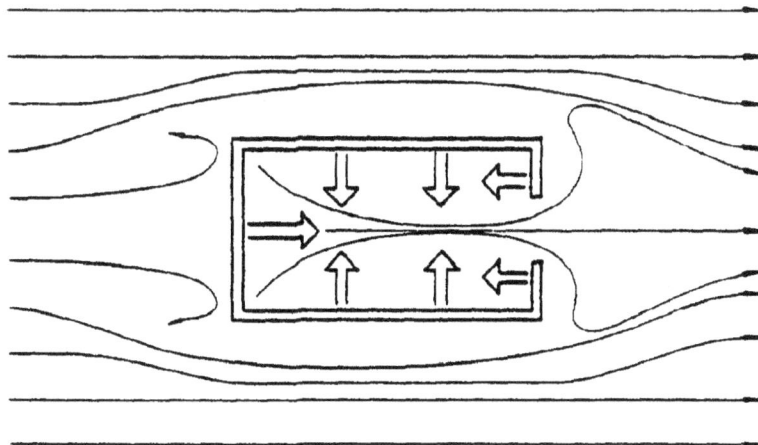

FIGURE 2-17
INTERNAL PRESSURE ON ROOF AS RESULT OF LEEWARD OPENING FAILURE

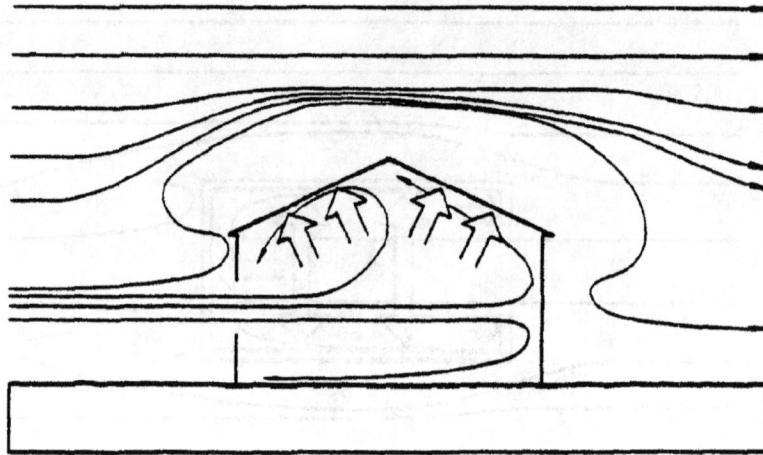

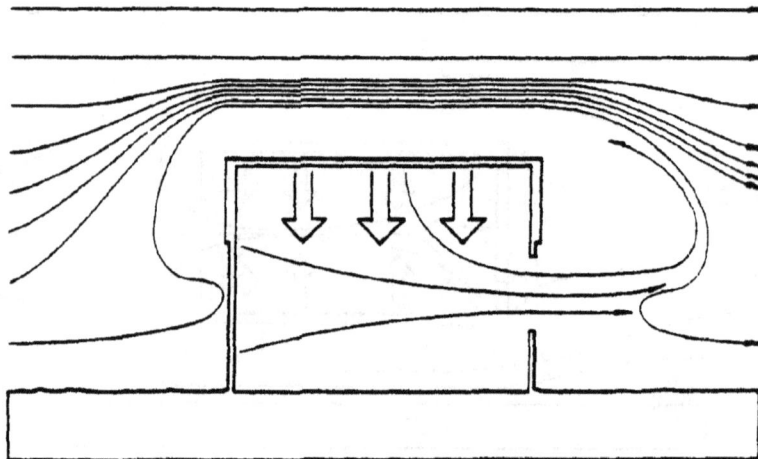

One of the common myths about hurricane resistance is that windows should be left open to equalize internal and external pressures. The myth is not borne out in experience. Where the failure occurs on the windward side and internal pressure increases, the increased outward pressure on the walls and roof can be as much as 1.6 times the pressure occurring when the entire building envelope remains intact. Houses that resisted Hurricane Alicia's highest wind forces well invariably had storm shuttered windows which, with the rest of the envelope, remained intact. During a hurricane pressure changes can occur rapidly, but generally not so rapidly that normal cracks and fissures in the building cannot equalize them before the building is damaged. Resistance to wind forces -- transmitting forces from the peak of the roof through the walls, deck and foundation into the ground -- and keeping the building envelope intact, are generally the most important factors in a building's survival.

D. **Combination of External and Internal Pressures** - When a door or window fails, external and internal wind pressures act simultaneously, and may increase the loadings on various components.

E. **Windward Wall, Window or Door Failure** - When the failure occurs in the windward wall, the outward external pressures on the roof, side walls and leeward walls and the internal pressures combine resulting in increased loadings on these components (Figures 2-18, 2-20 and 2-21). The load on the windward wall is decreased as the internal pressure acts opposite to the external pressure. Should the failure occur in the leeward or side walls, the inward external pressures and the internal pressure on the windward wall combine resulting in an increased loading on the windward wall (Figure 2-22). The loads on the other walls and roof are decreased as the internal pressure acts opposite to the external pressure (Figure 2-19, 2-23 and 2-24).

FIGURE 2-18
COMBINED ROOF LOADINGS - WINDWARD OPENING FAILURE

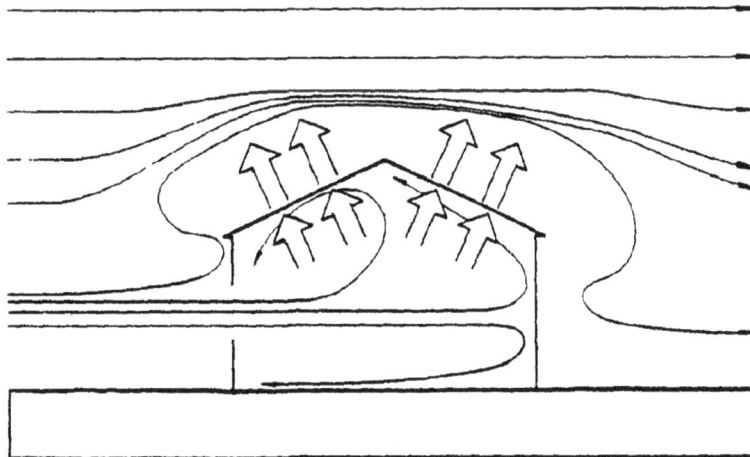

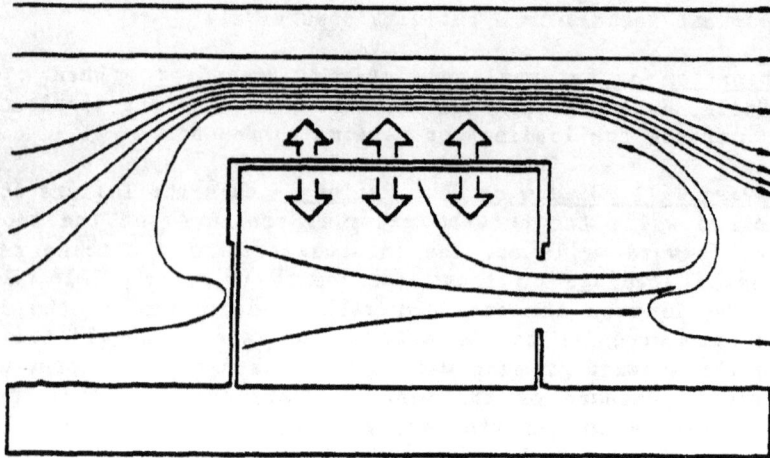

FIGURE 2-19
COMBINED LOADING ON ROOF AS RESULT OF LEEWARD OPENING FAILURE

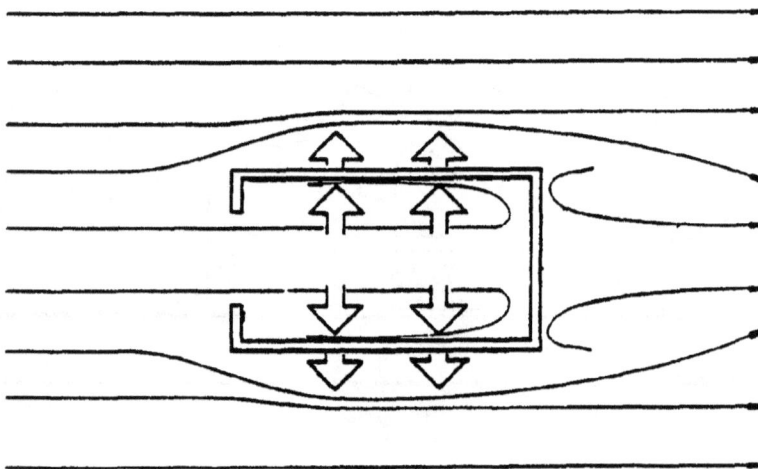

FIGURE 2-20
COMBINED SIDE WALL LOADINGS - WINDWARD OPENING FAILURE

FIGURE 2-21
COMBINED LEEWARD WALL LOADINGS - WINDWARD OPENING FAILURE

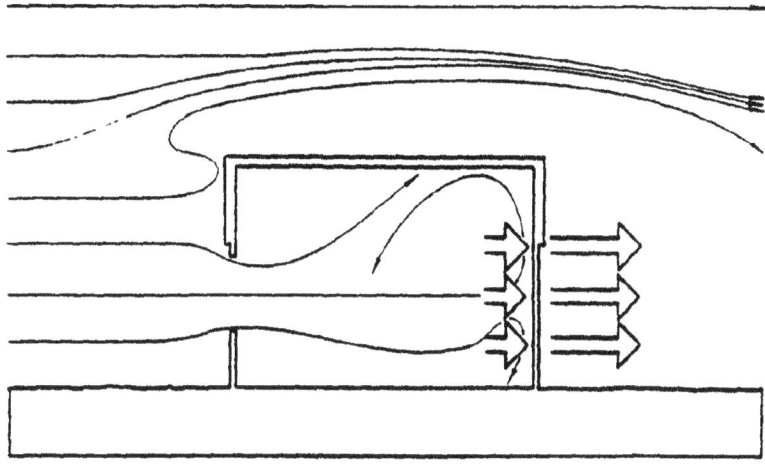

FIGURE 2-22
COMBINED WINDWARD WALL LOADINGS - LEEWARD OPENING FAILURE

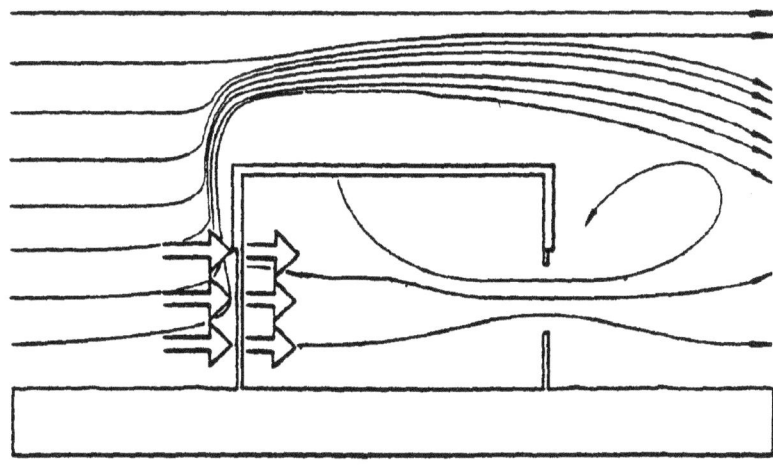

FIGURE 2-23
COMBINED WALL LOADINGS - LEEWARD OPENING FAILURE

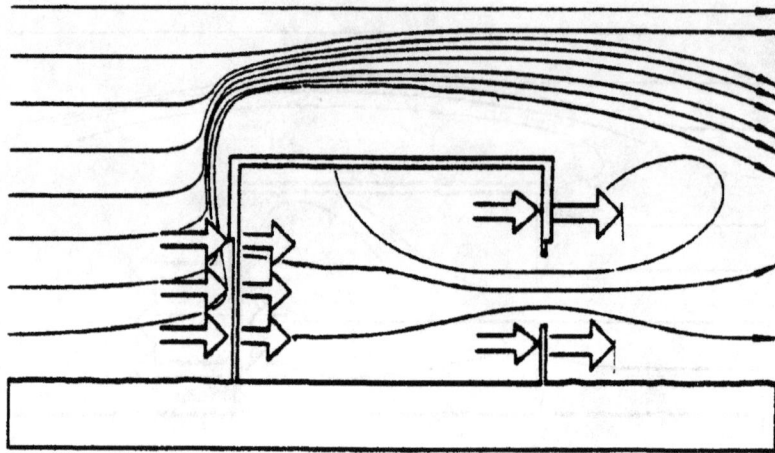

FIGURE 2-24
COMBINED WALL LOADINGS SIDEWALL OPENING FAILURE

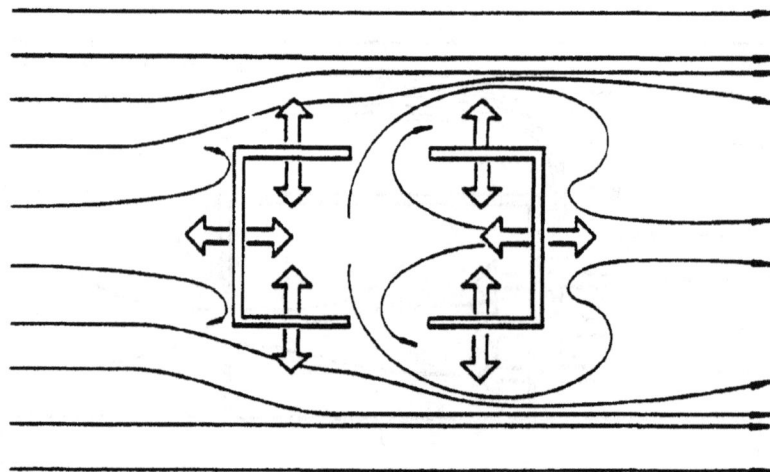

IV. COMBINED WATER AND WIND LOADINGS

When both water and wind interact with a building at the same time, the forces may combine and substantially increase the total loads the building or building components must resist. Figure 2-25 and 2-26 reflects two possible load combinations for wind and water.

FIGURE 2-25
COMBINED SURGE, WAVE AND WIND FORCES ON BUILDING

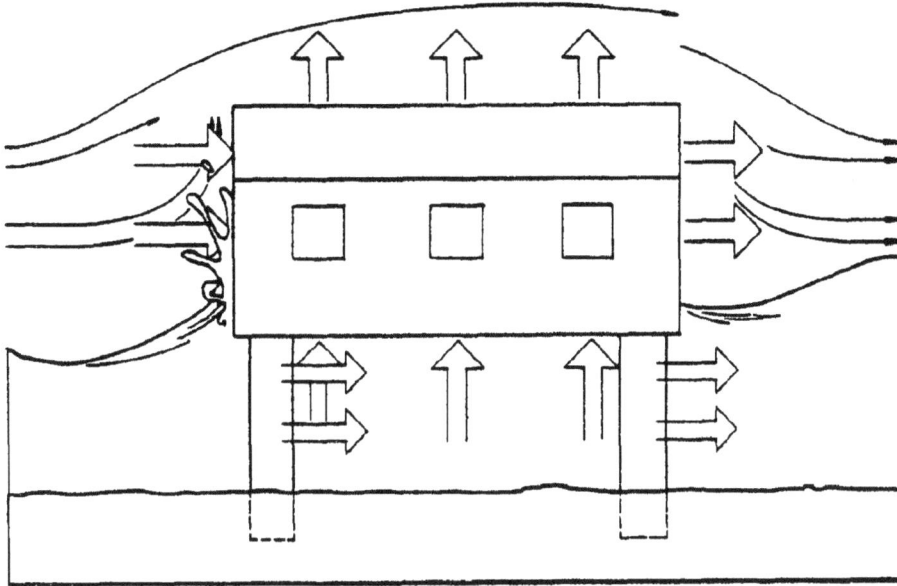

FIGURE 2-26
SURGE AND WAVE FORCES ON PILES AND WIND FORCES ON WALLS AND ROOFS

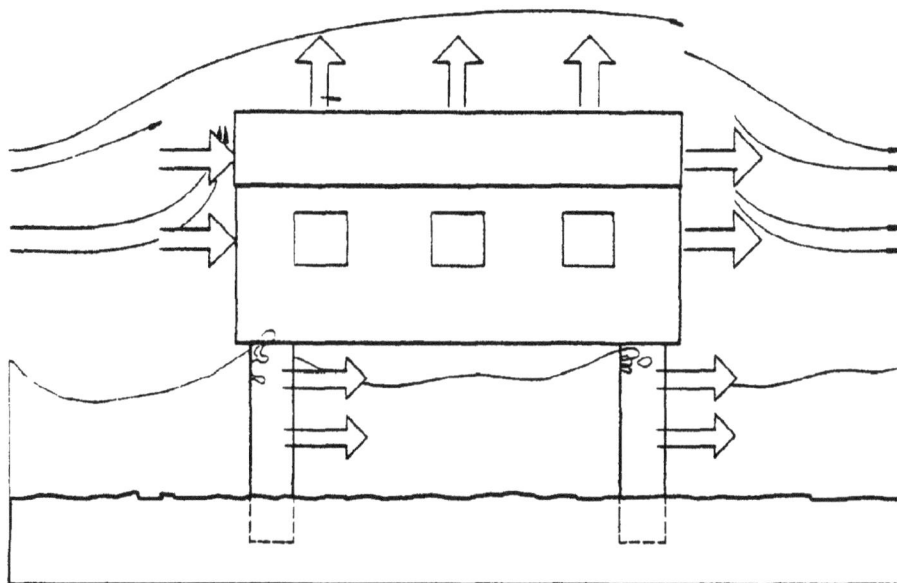

Hurricane forces interacting with buildings make coastal building design and construction complex and unique. Complex in that the building must be designed and constructed to resist both wind forces and water forces of the hurricane, unique in that close attention must be paid to all details. Lack of attention to a connector detail may well create a weak link in the structural chain, resulting in the failure of the building. The following chapters re-emphasize these points, and identify the most critical areas in the structural chain.

The most practical approach to coastal construction is to require that a registered architect or engineer design all buildings subject to hurricanes. Practically, for many of the one and two family dwellings constructed, a registered architect or engineer is not required. Recognizing this, Chapter IV "Construction Materials and Methods" was developed to provide the owner, builder and inspector with prescriptive requirements and details that would resist the forces of most hurricanes.

CHAPTER III
CONSTRUCTION DEFICIENCIES

INTRODUCTION

Special precautions must be taken in the design and construction of hurricane resistant buildings. The precautions start at the very base of the foundation and extend to the peak of the roof, tying the building together as an integral whole to resist the wind, water and battering of a hurricane. The need for the precautions can be seen in an aerial overview of an area struck by a hurricane, such as Galveston after Hurricane Alicia. There is ample evidence that lack of attention to proper details can result in the failure of one building while its next door neighbor suffers minor or no damage.

Figure 3-1. In the westernmost subdivision of Bay Harbor, near where the eye of Hurricane Alicia made landfall, we see a number of characteristics that persisted in subdivisions further to the right or east of the storm. In the foreground is about a two foot deep scarp from scour that in some locations was as deep as six feet. While about two feet of water covered the island at this location, sand was overwashed only to San Lulis Pass Road by the waves before settling out. While the house in the foreground seems to have suffered only minor roof damage, ground survey showed that extensive interior damage was caused when the windows broke and the east (right) wall separated at the floor plate. Buildings in this subdivision were older than most and little in their construction distinguished them from houses that might be built inland. One result was an unusually high amount of damage and destruction to homes in the middle foreground. Note, however, that there is relatively little damage evident to homes on the Bay side in the background.

FIGURE 3-1
AERIAL VIEW BAY HARBOUR, TEXAS

FIGURE 3-1
AERIAL VIEW BAY HARBOUR, TEXAS

Figure 3-2. Both the houses in the foreground suffered extensive damage. The superstructure of the house on the left failed at the floor plate when heavily corroded bolts embedded in the concrete deck failed to hold. The house on the right suffered greater than 50 percent damages to the roof, walls, decks, windows, and ground level enclosed area, while the pilings and grade slab remained intact.

FIGURE 3-2
AERIAL VIEW TWO HOUSES

Figure 3-3. Further east in the Terramar subdivision, houses were
built closer to the water than in Bay Harbor, and scour extended
beneath and beyond most of the first tier of houses. Where scour
undercut the grade slab, the slab typically failed, largely for
failure to tie the slab to the pilings and because reinforcing steel
mesh was laid directly on the ground before concrete was poured.
Noted the damages to the second and third houses from the left
foreground, while the two houses on the right fared well, with water
damages limited primarily to the grade slabs, enclosed ground level
area, bulkhead failure, and septic system failure. Almost all of the
houses in the first tier in this subdivision were condemned because of
septic system failure.

FIGURE 3-3
AERIAL VIEW TERRAMAR 1 SUBDIVISION

Figure 3-4. Both success and failures are evident. The house on the upper left incurred only roof damage and failure of a sliding glass door, while the remaining houses incurred major damage and 3 were virtually destroyed. The house in the lower center lost its entire superstructure at the floor plate where nails (no ties) were inadequate to resist the wind pressures. The house upper right center appears to have incurred only minor damage, however, it was gutted as a result of window and door failures on the wall facing the gulf.

FIGURE 3-4
SUCCESS AND FAILURE

By now, patterns of damages and destruction are well established. Typically, houses in the first tier from the Gulf received the greatest damages from water and wind. The greater the setback from the Gulf, the less likely that the house received flooding damages (Figure 3-1). However, wind damages were experienced across the island, although the farther inland, the less likely the house was extensively damaged, with notable exceptions.

Failure can take many forms. There can be a structural collapse (Figure 3-5).

FIGURE 3-5
STRUCTURAL COLLAPSE

Part of the roof can blow off resulting in water damage to the building interior (Figures 3-6).

FIGURE 3-6
PARTIAL ROOF FAILURE

Windows can be broken by high winds or debris impact, resulting again in water damage or magnified wind damage (Figures 3-7).

FIGURE 3-7
WINDOW FAILURE

Whether by wind or water, unreinforced masonry walls are a hazard both structurally (Figure 3-8) and by the emotions they evoke. Brick veneer and other cladding without ties often collapses under wind pressures (Figures 3-9).

FIGURE 3-8.
UNREINFORCED MASONRY FAILURE

FIGURE 3-9
BRICK VENEER FAILURE

There can even be progressive corrosion resulting in structural failure, such as the corroded floor bolts that failed on several houses in Galveston.

Any building condition resulting in bodily harm or property damage can be termed a failure.

II. PILE FOUNDATIONS

A pile foundation takes on extra significance in coastal flood plains exposed to hurricanes. Not only must the foundation support the superstructure, but it must also support the lateral and uplift loads applied to the superstructure.

Depth of embedment, illustrated below, is critical in areas subject to storm surge and scour. Scouring of soil can result in insufficient friction resistance to uplift and lateral forces. Even minimal scouring combined with pile installation by jetting or auger can result in the same failures, because the soil may be insufficiently compacted to provide essential friction resistance. Granular (sandy) soils are particularly susceptible to scouring, and inadequate embedment can leave a house hanging (Figures 3-10) or collapsed (Figures 3-11).

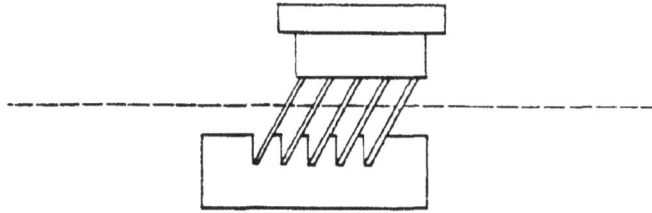

FIGURE 3-10
INADEQUATE PILE EMBEDMENT DEPTH

FIGURE 3-11
BUILDING COLLAPSE AS RESULT OF INADEQUATE PILE EMBEDMENT

Undersizing of piles will result when wind loads or scouring are underestimated. An undersized pile properly embedded will fail by bending at the scoured soil line (Figure 3-12), or failing catastrophically (Figure 3-13). Without scour, but underdesigned for wind, the bending failure can occur (Figure 3-14). Lack of proper pressure treatment of wood piles can result in undersized piles through decay.

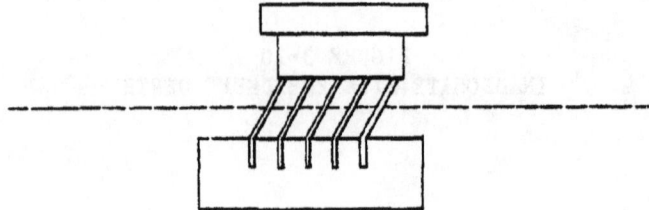

FIGURE 3-12
PILE FAILURE AT SCOUR LINE

FIGURE 3-13
CATASTROPHIC PILE FAILURE

FIGURE 3-14
BENDING FAILURE

Cross-bracing of piles, illustrated below, can create unforeseen
problems due to entrapment of debris in the bracing members, which
creates a larger surface area for wind and waves to act on.
Frequently the owner will remove part of the cross bracing to allow a
clear unobstructed access (Figure 3-15), thus compromising the
structural integrity of the cross bracing.

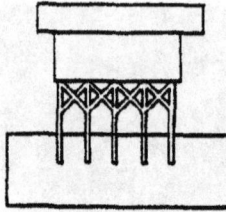

FIGURE 3-15
CROSS BRACING AND GROUND FLOOR SLAB

III. GROUND SLAB

Often the space under the elevated first floor of a beach house, illustrated below, is used for parking and storage. A concrete ground slab in this space is subject to severe undercutting by scouring action of the storm surge and waves (Figure 3-15). An undercut slab will fail unless tied to the piles and designed to span between piles. If it is tied to the slab and scouring action occurs, the weight of the slab is supported by the piles.

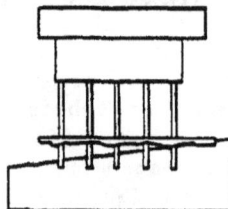

IV.   BREAKAWAY WALLS AND STAIRS

Rooms enclosed areas and equipment located below the wave crest level
and subject to wave action during a storm, illustrated below, can be
considered dedicated to the storm.   Evidence suggests that however
built, such enclosed areas tend to be destroyed or heavily damaged,
even when the rest of the building may survive the storm relatively
intact.   And if not subject to wave action but to storm surge,
contents in such areas are typically damaged by the rising floor
waters.   Sound reasons why the National Flood Insurance Program
refuses to issue flood insurance on the structure and contents below
the 100-year flood level in coastal homes.

Where areas below the 100-year flood level are to be enclosed, walls
and partitions should be designed with light weight or nonfloating
materials, and with minimal connections to the first floor structure
and pilings (Figure 3-16).   This allows the initial water impact to
collapse the walls and stairs before significant forces are
transmitted to the habitable structure and its foundations, and the
resulting debris is light or will sink without increasing lateral
loads on the house or nearby buildings.

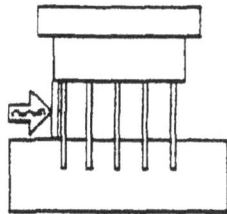

FIGURE 3-16
BREAKAWAY WALL

V.   CONNECTIONS

Improper connections are a major contributor to structure damage in a hurricane. They come in many forms: the top of an undersized piling further weakened by cutting over half the piling for a resting surface for a beam; an undersized rebar inadequate to resist wind and wave forces; and, most frequently, no special connections at all. Special connections must form a continuous tie from the roof to the foundation.

The coastal environment is extraordinarily corrosive. Connections must also be sufficiently corrosion resistant to provide adequate long term protection, or else they will be substantially weakened by the corrosion, and may fail when put to the test by a storm.

The illustration below shows that high wind forces can blow the entire superstructure off the foundation without adequate connection between top of piles and floor framing (Figure 3-17).

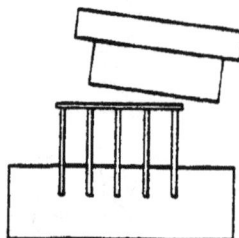

FIGURE 3-17
INADEQUATE CONNECTIONS

Beams must be sturdily fastened to the pilings and beams to joists. Studs must be tied to the bottom plate, which in turn must be tied to supporting floor beams. Failure to make a strong, secure connection can result in the superstructure separating from the floor (Figure 3-17). Similarly, studs must be tied to the top plates and rafters, illustrated below, or the roof will separate from the remaining structure (Figure 3-18). A variety of galvanized hangers and framing anchors are widely available on the market and are appropriate for coastal construction (Figure 3-19).

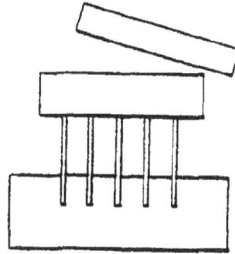

FIGURE 3-18
ROOF CONNECTOR FAILURE

FIGURE 3-19
CONNECTOR MANUFACTURERS

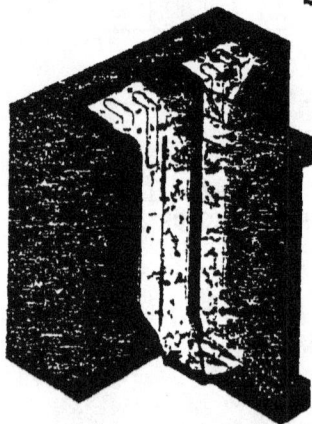

VI. OPENINGS

Openings, particularly window openings, illustrated below, are often the weakest link in a storm resistant structure. A broken window can cause significant increases in internal pressures and magnify the chances of further structural damage. At the least, water damage can be expected when the building envelope is ruptured. During Helena, plywood storm shutters proved effective in protecting windows (Figure 3-20).

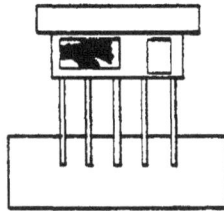

FIGURE 3-20
PLYWOOD SHUTTERS

Even small openings, such as those found in a loosely constructed building, can cause a sufficient amount of internal pressure to cause unexpected structural damage.

VII. ROOF CONSTRUCTION

Integrity of roof sheathing, illustrated below, is essential for protection against wind and water. A breach in the sheathing, besides allowing entrance of rain water, can also start a chain reaction through pressure buildup in the attic space ending in complete destruction of the roof structure (Figure 3-21).

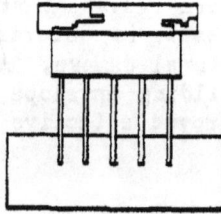

FIGURE 3-21
PROGRESSIVE ROOF FAILURE COLLAPSE

Failure at the gable end of a pitched roof, illustrated below, has the
same effect as a sheathing failure. Gable ends covered only with thin
wall covering can blow in. When covered with adequate wall materials
but without lateral bracing of rafters or trusses, the gable end can
suffer local collapse (Figures 3-22).

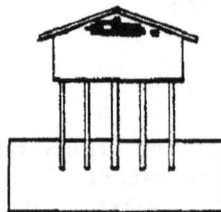

FIGURE 3-22
ROOF SHEATHING FAILURE

VIII. SUMMARY

This chapter has emphasized the weak link failure mechanism. A failure in one or more of the structural components may well result in the failure of one or more additional components resulting in many cases in catastrophical failure of the building structural system and collapse of the building (Figure 3-23). If attention is given to both the design construction and inspection, Chapters IV and V, your home may well resist the forces of a hurricane (Figure 3-24).

FIGURE 3-23
TYPICAL RESIDENTIAL BUILDING FAILURE

FIGURE 3-24
DAUPHINE ISLAND RESIDENCE THAT INCURRED
MINIMAL DAMAGE IN HURRICANE HELENA

CHAPTER IV

CONSTRUCTION MATERIALS AND METHODS

I.   INTRODUCTION

It is not the intent of this Chapter to minimize the importance that
a registered architect or engineer design all buildings subject to
the forces of hurricanes. The intent is to provide prescriptive
design information for those areas in which state or local laws do
not require a registered architect or engineer to design one and two
family dwellings and small commercial buildings.

Designs by registered architects or engineers, will result in
requirements that differ from those presented in this Chapter. The
primary reason for the difference is the architect or engineer is
working with site specific design data. Where a set of plans and
specifications have been developed for a one and two family dwelling
by a registered architect or engineer, it is recommended that these
documents be processed for permit issuance in the same manner as any
other architect or engineer designed building.

II. MATERIALS

There are many materials used in the construction of buildings. In
coastal construction, certain of these materials may not afford the
same advantage as they afford in noncoastal construction. The
purpose of this chapter is to present these various structural
materials in the way they are used in constructing a building and
any special requirements and advantages or disadvantages, if any,
for coastal construction.

Most structural construction materials are aluminum, concrete,
masonry, steel or wood.

A. Aluminum materials are used for window and door assemblies,
gutters, downspouts and flashing. Because of the rapid
deterioration in the coastal environment, special finishes such as
an anodizing or vinyl finish is necessary. Care must be employed
with the protective finish in the installation of aluminum
components to insure that the finish remains intact.

B. Concrete materials are used for pier pilings, beams, walls and
floor panels, etc. This material performs well in the coastal
environment. The major concern is that the steel reinforcement must
be covered with sufficient concrete to reduce the danger of salt
water reaching the steel and causing it to corrode. This concrete
coverage varies from 3" for concrete cast against and permanently
exposed to earth (portion of pile in ground) to 3/4" for some
precast plant controlled wall panels. Concrete exposed to salt
water spray requires additional precautions, such as increased
minimum strength and use of sulfate resisting cements. Concrete
Piles for one and two family dwelling construction is normally more
expensive than wood. This cost difference begins to equal out when
the residence is elevated to approximately 16 feet and the total
construction costs exceeds $100,000.00.

C. Masonry materials are used primarily for walls and piers where spread footings are used. In elevated structures, masonry construction is not practical because of the special nature of the connections and the increased dead load.

D. Steel materials are used for piles, beams, walls and floor panels, siding, roofing and connectors. Steel is subject to corrosion. A recent test indicated that rusting occurred six times faster 80 feet inland than a site 800 feet inland. The corrosion rate reaches a maximum at approximately 10 feet off the ground and then decreases. The fastest corrosion occurs when the steel is partially enclosed. The partial cover allows the salt spray to accumulate and prevents periodic washing by rainfall. The salt becomes more concentrated on the steel and stays damp longer; the area subject to the highest corrosion rate is the underside of an elevated beachfront building. Areas within walls or under roofs are not free from potential corrosion. The air on the coast contains significant amounts of salt moisture and as a result of the movement of air through these areas, salt and moisture are deposited. This means nails and other connectors within walls, ceiling, etc. are subject to corrosion.

The primary way to reduce corrosion is by galvanizing the steel. Proper galvanizing can reduce the corrosion by a factor of 20. All steel used in coastal areas should be galvanized to form a zinc coating of at least 1 oz. per square foot, including bolts, nails and connectors. Zinc may be applied by hot-dip galvanizing, electrogalvanizing or mechanically deposited. Hot-dipped galvanizing has the advantage of depositing the heaviest coating.

E. Wood is the most available and commonly used material for one and two family dwellings on the coast. With proper selection, spacing and sizing, wood is a very satisfactory construction material for one and two family dwellings. Wood in contact with both air and water or the ground is subject to decay. Wood also placed in salt water is subject to destruction by marine borers. Wood members in contact with the ground and above the waterline need only to be naturally resistant to decay (e.g., heartwood of cedars, redwood) or be pressure preservative treated in accordance with standards of the American Wood Preservers Bureau or the American Wood Preservers Association.

There are two species of wood that are most often pressure-treated in the United States, Southern Pine (Loblolly, Slash, Shortleaf and Longleaf) and Coastal Douglas Fir. In the Southern and Eastern States, treated Southern Pine is usually supplied, however, because of pine's size limitations, piling over 75 feet long are often furnished in Douglas Fir.

Three types of preservations are available for pressure-treatment: (1) creosote; (2) pentachlorophenol (penta); and (3) waterborne [Chromated Copper Arsenate (CCA), Ammoniacal Copper Arsenate (ACA), and Ammoniacal Copper Zinc Arsenate (ACZA)]. Where waterborne is used, Southern Pine is usually treated with CCA, ACA or ACZA. Penta cannot be used in saltwater applications.

Creosote is generally the preservative used for foundation piling and CCA for construction that may be subject to contact by people - cleanliness being the deciding factor In marine waters, creosote or creosote-coal tar solution treatments are used in Northern water where toredo (shipworm) and pholad attack are known or expected and where limnoria tripunctata attacks are not prevalent; two, CCA or ACA is used where toredo and limnoria tripunctata attack are known or expected and pholad attack is not prevalent; and three, dual-treatment of CCA or ACA and creosote are used in areas where limnoria tripunctata and pholad attacks are known or expected, i.e., the warm Southern waters.

Wood treated in accordance with the American Wood Preservers Bureau and the American Wood Preservers Association standards are labeled. The presence of this label is the best way the builder or inspector can be assured that piles have been preservative treated. Each pile shall be clearly and permanently branded at 5 feet and 10 feet from the butt. Branding may be accomplished by "burn branding" or attaching a metal disk in a pre-bored recess. Metal disks shall be of 20 to 24 gauge sheet monel metal and all lettering and numbers shall be not less than 1/8 inch high. The brand shall contain the following information:

1. Supplier's brand.
2. Plant designation, month and year treated.
3. Species of timber, length, and (from ASTM D25) Table 1 or 2.
4. Type of preservative.
5. Retention

See Figure 4-1 for an example of typical band and key. The retention of preservative in pounds per cubic foot is dependent upon the species of wood, the material the pile is driven into and the preservative. Table 4-1 indicates these requirements for Southern Pine and Douglas Fir piles.

FIGURE 4-1
TYPICAL WOOD PRESERVATIVE BRAND AND KEY

ABCO	------------ Supplier's Brand
D	------------ Plant Designation
60	------------ Year of Treatment
SPC	------------ Species of Treatment and Preservative Treatment
---	------------ Retentions
7-30	------------ Class and Length

Source - American Wood Association Standard M6-75, Brands Used on Forest Products

TABLE 4-1
PRESERVATIVE RETENTION REQUIREMENTS

Preservative	Southern Pine	Douglas Fir

Piles subject to salt water and wave action

Creosote	20.00 pcf	20.00 pcf
*Penta	Not Recommended	
*ACA	2.50 pcf	2.50 pcf
*CCA and ACZA	2.50 pcf	2.50 pcf

Piles not subject to salt water or wave action

Cresote	12.00 pcf	12.00 pcf
Penta	.60 pcf	.85 pcf
*ACA	.80 pcf	1.00 pcf
*CCA and ACZA	.80 pcf	1.00 pcf

*Penta - Pentachlorophenol
*ACA - Ammoniacal Copper Arsenate
*CCA - Chromated Copper Arsenate
*ACZA - Ammoniacal Copper Zinc Arsenate

The AWPA Quality Mark incorporating the legend "MP-1, MP-2 or MP-4" shall be stamped on a monel metal disk which shall be the Quality Mark. The Quality Mark shall be permanently affixed to each treated pile with a monel nail in a depression on the side of the pile between the two burned brands. When this Quality Mark is attached to the pile, it certifies that all requirements under American Wood Preservative Standard MP-1, MP-2 or MP-4 have been met by both the treater and control agency. An example of a typical label is shown in Figure 4-2.

FIGURE 4-2
EXAMPLE WOOD PRESERVATIVE TREATMENT LABELS

Source - Society of American Wood Preserver's Inc.

III. FOUNDATIONS

A. Introduction

Materials for foundations are dependent upon the type foundations. If the structure is not elevated and soil conditions permit, the foundation is normally a spread footing (Figure 4-3) which transmits the load of the building into the ground. This footing can be constructed of concrete or pressure treated wood. When the building is to be elevated or the soil conditions will not support the load, piles (Figure 4-4) are used. Piles may be wood, concrete or steel.

FIGURE 4-3
TYPICAL SPREAD FOOTING DETAILS

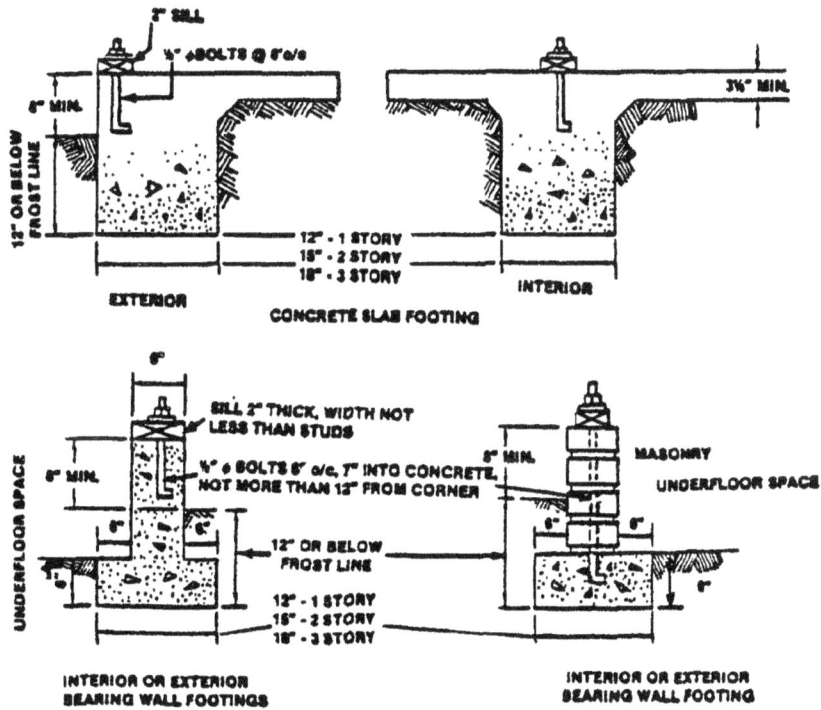

Source - 1983 Edition, CABO One and Two Family Dwelling Code

FIGURE 4-4
TYPICAL PILE FOUNDATION

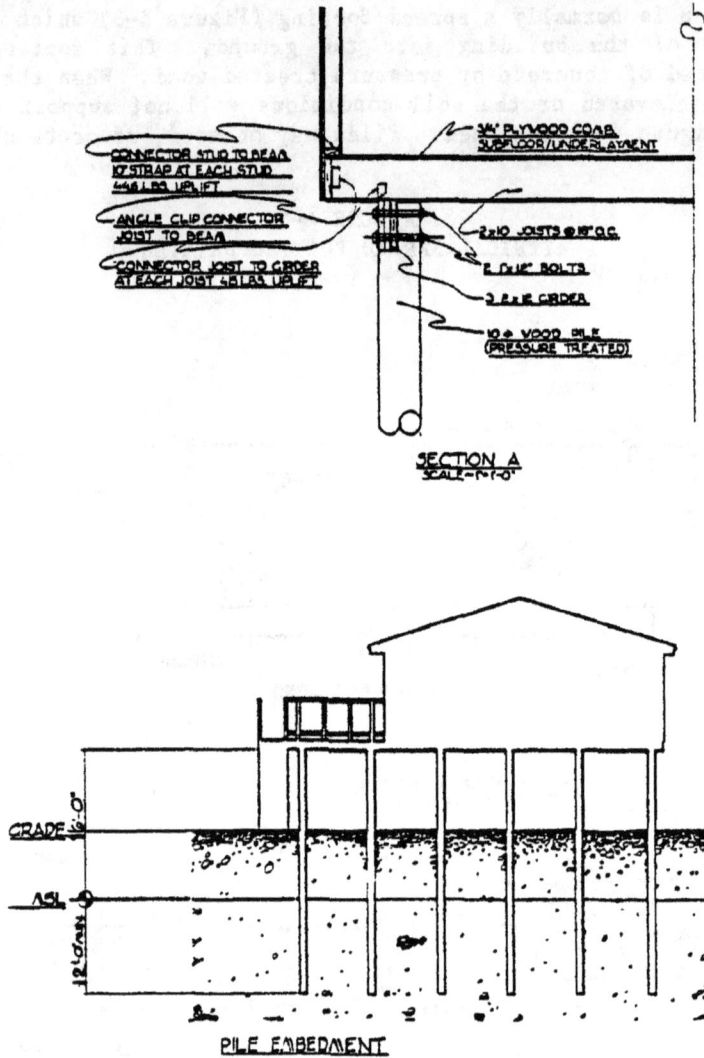

SECTION A
SCALE~1'-1'-0"

PILE EMBEDMENT

B. Spread Footings

In coastal areas spread footing foundations should be used only in areas not subject to storm surge and wave action. This type foundation is used in construction throughout the United States. The reader is advised that the One and Two Family Dwelling Code published by the Council of American Building Officials contains requirements for these foundations. When foundation walls and spread footings are used in flood areas, openings should be provided

in each foundation wall face to allow equal action of water pressure
on the foundation walls. In nonflooding areas, building failure due
to hurricane winds generally has been the result of inadequate
connections between the foundation and the building walls. The
recommended details for these connections will be discussed later.

C. Foundation Wall, Post and Piers

Foundation walls, piers, and posts support the structure above and
transmit the load to the spread footings. Since they attach to
spread footings, foundation walls, posts and piers are not
recommended for coastal areas subject to flooding. Typical details
of foundation wall systems are shown in Figure 4-3. Typical post
and pier foundations are shown in Figure 4-5.

FIGURE 4-5
TYPICAL POST OR PIER

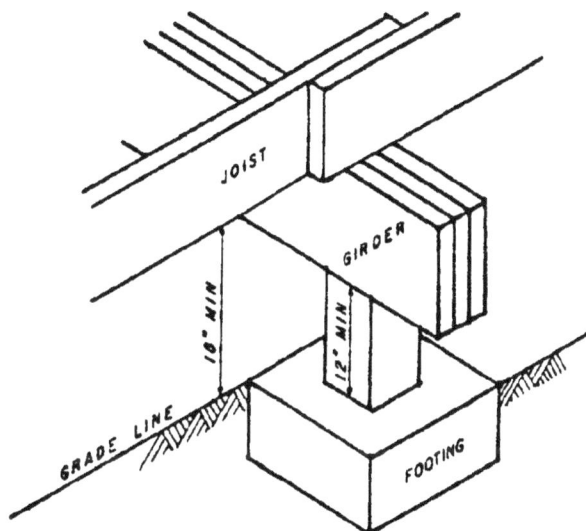

D. Piles

Piles are concrete, steel or wood. Wood is the most commonly used
material for piles in one and two family dwellings because of the
availability of the material, use of standard connectors and costs.
When residences are elevated to heights of approximately 16 feet,
the cost differential between wood piles and beams and precast
concrete piles and beams may be minimal. Since this manual is based
on one and two family dwellings, the further presentation on piles
will be limited to wood.

Builders prefer square timber piles because of the ease of framing
and the connection of the building structural system to the piles.
Two sizes, 10" x 10" and 8" x 8" rough-sawn members are the most
frequent used piles. Round piles are generally stronger, longer and
cost less than square piles. When round piles are used, a minimum
tip diameter of 8" is the generally accepted minimum.

Piles must resist both vertical loads and horizontal forces.
Vertical gravity loads are the weights of the structure, the
occupants and the furnishings. Horizontal forces include the wind
acting on the walls and roof (Figure 4-6). Wind may cause upward
and downward forces by transmitting forces through walls, floor and
roof to the piles.

Lateral forces (horizontal) can be produced by both wind and water
(Figure 4-7). Wind acts primarily on the walls, water produces drag
forces on the piles, and water impacts on any portions of the
structure below the water level to create lateral forces. When wave
action accompanies the storm surge, the soil around the pile is
subject to erosion (scour).

FIGURE 4-6
WIND FORCE

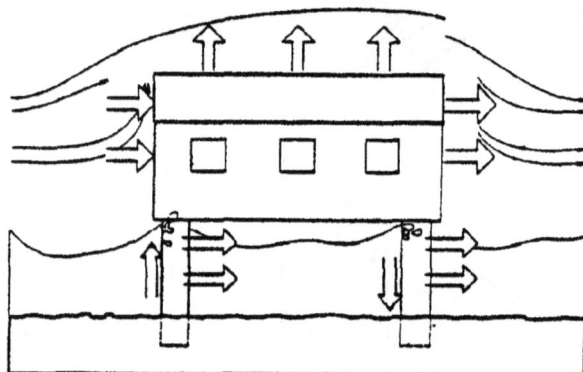

FIGURE 4-7
LATERAL FORCE

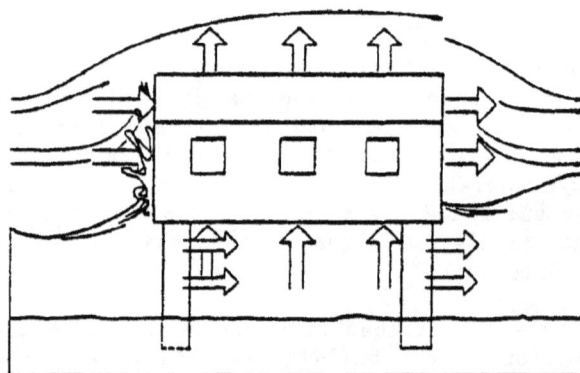

1. Pile Installation

   The effectiveness of pile foundations is largely determined by
   the method by which piles are installed in the ground. There
   are three methods used to place piles: driving, excavating and
   jetting.

   (a) Driving

   Driving a pile into the ground is the most effective. Driving
   forces the soil out from the pile butt, makes the surrounding
   soil more dense, and thus increases the friction along the
   surface of the pile. This increase in friction allows the pile
   to support greater loads.

   Pile driving is accomplished through the use of pile hammers.
   Pile hammers are devices that impart energy to the pile causing
   the pile to penetrate the soil. There are normally three types
   of pile hammers used for residential and small commercial
   buildings.

      (1) Drop hammers are used for small, relatively
      inaccessible, jobs. The drop hammer consists of a metal
      weight fitted with a lifting hook. The hook is connected
      to a cable which fits over a sheave block and is connected
      to a hoisting drum. The weight is lifted and tripped,
      freely falling to a collision with the pile. The impact
      drives the pile into the ground. The major disadvantages
      are the slow rate of blows and length of leads required
      during the early driving to obtain a sufficient height of
      fall to drive the pile.

      (2) Single-acting hammers use pneumatic pressure to lift
      the ram to the necessary height. The ram then drops by
      gravity onto the anvil, which transmits the impact energy
      to the capblock, and then to the pile. The hammer blows
      are delivered at a relative rate. The hammer length must
      be long to obtain a impact velocity (or height of ram
      fall), that produces sufficient energy to drive the pile.
      The blow rate is considerably higher than that of the drop
      hammer. In general the ratio of ram weight to pile weight
      including appurtenances should be on the order of between
      0.5 and 1.0.

      (3) Double acting hammers use pneumatic pressure to lift
      the ram and to accelerate it downward. The blow rate and
      energy output is usually higher for double-acting hammers,
      but steam consumption is also higher than for the
      single-acting hammer. The hammer length may be several
      feet shorter for the double-acting hammer than for the
      single-acting hammer. The ratio of ram weight to pile
      weight should be on the order of between 0.5 and 1.0.

   (b) Excavating

   A hole bored with an auger is occasionally used in clay or silty
   soils. The auger is removed and the pile is placed into the
   hole. Sand, gravel or similar materials are poured and tamped
   into the hole around the pile. Frequently the compaction of the

sand or gravel around the pile is not sufficient to provide a
lateral support comparable to driven piles. In order to
increase the support a combination excavation and driving is
used to place the piles. Normally the piles are driven for the
last 4 to 8 feet. These methods are generally used only for
bearing piles since the friction resistance between the pile
and the soil is minimum.

(c) Jetting

A too frequently used method for inserting piles into sand is
called jetting. Jetting is a method by which a pipe alongside
the pile concentrates a high pressure stream of water below the
pile. The water stream continuously pushes a hole in the sand
below the pile, with the pile weight advancing the pile to the
desired depth. Sand or gravel is then tamped into the cavity
around the pile. Jetting loosens the sand around and below the
pile and thus reduces both the friction and bearing resistance
of the soil. If this method is employed, the embedment depth
should be based on loose sand and load tests should be required.

2. Pile Sizes

In sizing piles and determining the depth of penetration, the
forces of wind and water, gravity loads, type soils, pile
inserting method and potential erosion must be accounted for in
the design. Tables 4-2 and 4-3 were developed based on a
consideration of these variables. Table 4-2 is based on sandy
soil and Table 4-3 on clay soil.

All pile sizes for structures located in areas subject to storm
surges and wave action, i.e., the V-Zone, are based on a pile
penetration of 12 feet below mean sea level, with the first
floor level 16 feet above the mean sea level (See Figure 4-8).
Size of piles for structures not subject to surge wave action or
erosion are based on a penetration depth of 12 feet below grade,
with the first floor elevation 8 feet above grade (See Figure
4-8). In addition, pile sizes are based on a minimum of 4 piles
in any one row.

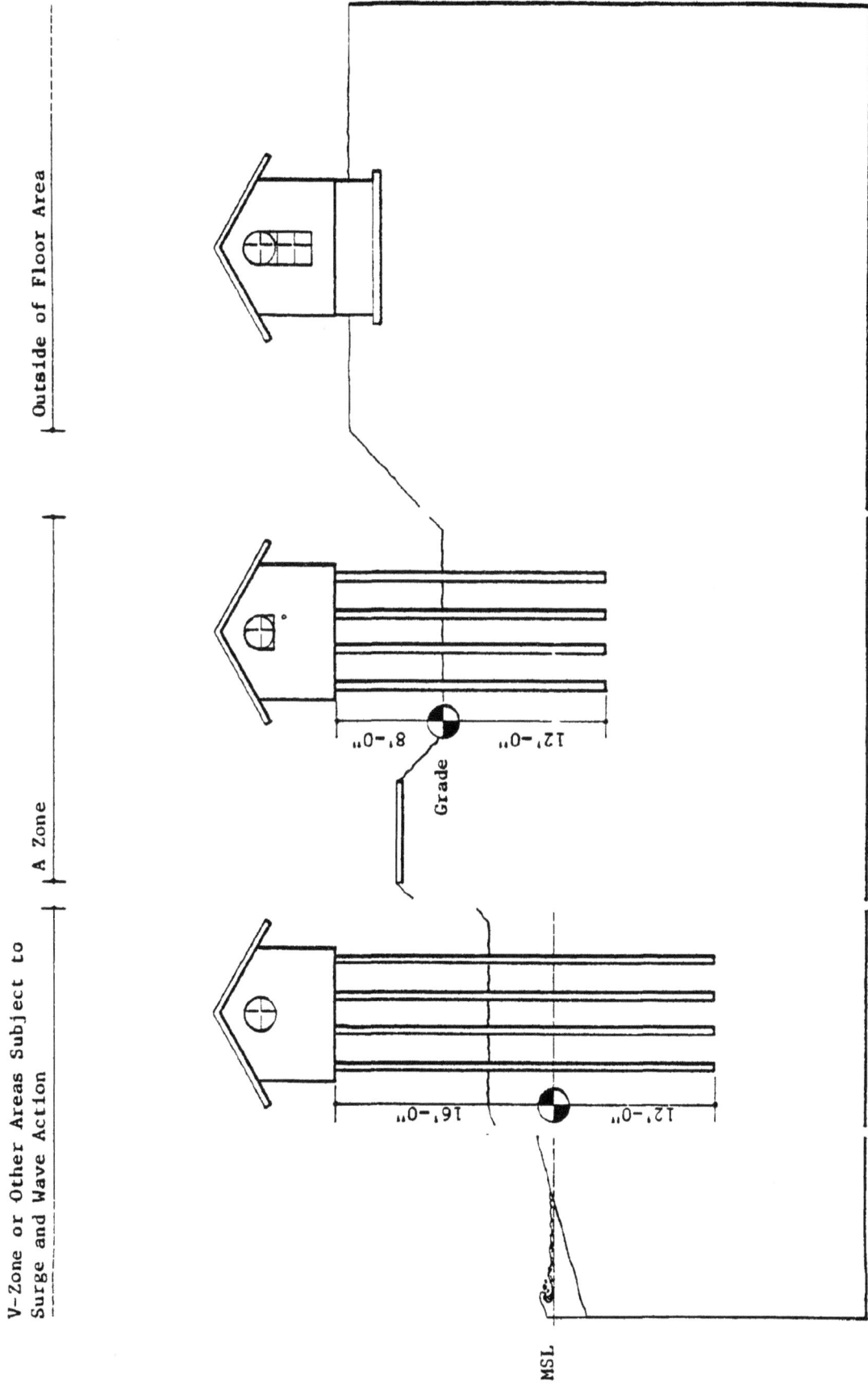

FIGURE 4-8
DEIGN ASSUMPTIONS

TABLE 4-2[2,3,4,5]
MINIMUM PILE DIAMETERS FOR SANDY SOIL

	110 mph maximum		120 mph maximum	
	Subject to Surge and Wave [1]	Not Subject to Surge and Wave [2]	Subject to Surge and Wave [1]	Not Subject to Surge and Wave [2]
	8'-0" Maximum Span			
one story	10" diam. tip	8" diam. tip	10" diam. tip	8" diam. tip
two story	10" diam. tip	8" diam. tip	12" diam. tip	10" diam. tip
	10'-0" Maximum Span			
one story	Not applicable	8" diam. tip	Not applicable	8" diam. tip
two story	Not applicable	10" diam. tip	Not applicable	10" diam. tip

Notes:

1. Minimum pile penetration 12 feet below MSL
   Maximum first floor elevation 16 feet above MSL
2. Minimum pile penetration 12 feet below grade
   Maximum first floor elevation 8 feet above grade
3. Pile size based on Southern Pine pile or Douglas Fir
4. Southern Pine or Douglas Fir square piles having the least dimension equal to the pile diameters shown in Table 4-2 may be substituted for round piles.
5. Minimum number piles in a row - 4.

TABLE 4-3[2,3,4,5]
MINIMUM PILE DIAMETERS AND PENETRATION DEPTHS FOR CLAY SOILS

	110 mph maximum	120 mph maximum
	8'-0" Maximum Span	
one story	10" pile @ 12' deep	10" pile @ 12' deep
two story	10" pile @ 20' deep	10" pile @ 20' deep
	10'-0" Maximum Span	
one story	10" pile @ 20' deep	10" pile @ 20' deep
two story	12" pile @ 24' deep	12" pile @ 24' deep

Notes:

1. Maximum first floor elevation 8 feet above grade
2. Pile size based on Southern Pine pile or Douglas Fir
3. Southern Pine or Douglas Fir square piles having the least dimension equal to the pile diameters shown in Table 4-2 may be substituted for round piles.
4. Minimum number piles in a row - 4.
5. Table 4-3 not applicable to areas subject to surge and wave.

IV.   OTHER BUILDING COMPONENTS BELOW SURGE AND WAVE LEVEL

A.   Slabs on Grade

It has been past practice in many coastal areas to pave the area
below the living area with concrete and use the area for parking
automobiles. This concrete grade slab also serves to restrain the
piles from moving at ground level, reducing the pile span by
approximately 2'-0", and allowing for a reduction in pile size.
Coastal areas surveyed that had surge and wave action indicate that
the soil under the grade slab typically washes out and does little
to restrain the piles. When this occurs, the slab only adds
additional load to the piles and makes soil replacement difficult.
If slabs are used in areas subject to storm surge and waves, the
slabs should be constructed so as to prevent the slab from
introducing any loads to the piling from erosion or scour under the
slab. This is normally accomplished by separating the slab from the
pile by an expansion joint and filling the joint with joint
expansion material.

B.   Cross Bracing

In order to reduce the impact of lateral forces on the piles and
possibly reduce the size or spacing of piles, cross bracing is
installed between piles. Cross bracing can be in the form of cable,
rods, or lumber (2x6, 2x8, etc.). The members have small surface
areas and offer little resistance to water movement. However, since
they may be below water during a storm, they are subject to the
impact of debris. If the debris is entrapped in the bracing, the
debris and bracing can provide significant resistance to water flow.
Any impact on flow resistance would be transferred as additional
concentrated forces by the bracing to the piling. In addition, the
owner frequently removes the cross bracing and compromises the
purpose of the bracing. For these reasons, cross bracing is not
recommended, and the piling sizes and spacings in this manual are
based on designs without cross bracing. If cross bracing is used,
it is recommended that it consist of cable or rod since cables or
rods offer the least resistance to waterflow and the smallest
surface area for attracting debris. In addition, if cross bracing
is used, continuous top and bottom bracing is required to all piles.
Partial bracing compromises the cross bracing system as the building
must be designed and constructed to resist wind loads from any
direction.

C.   Grade Beams

In some cases, a grade beam is installed between all piles at ground
level. The purpose of the beam is similar to the purpose of the
concrete slab - fixing the piles at ground level. Where erosion
occurs, these beams not only add additional vertical load to the
pile, but may also receive the impact load from debris and increase
resistance to water movement as the result of debris entrapment.
For these reasons, grade beams are not recommended and were not
considered in the design of pile sizes in this manual.

D. Walls and Partitions

The use of the area below the first floor of an elevated structure should be limited to uses that require no or limited walls and partitions. Any wall or partition structurally attached to the piles or floor above will increase the load from both wind and water on the building structural system. For this reason a wall or partition erected below the first floor should be of a type that will break away when subjected to wind or water loads. Breakaway walls or partitions of solid construction (e.g., plywood sheeting and wood studs) are not recommended because the wall or partition normally breaks away in large sections. These sections have the potential of hitting piles or becoming entrapped on cross bracing, significantly increasing the lateral force on the piling of the same or other buildings. The recommended construction for walls and partitions in this area is light lattice work or screens. This type construction would produce light debris that would create minimum impact loads on the building structural system.

Table 4-4 provides the maximum recommended design loads for breakaway walls. The design using the loads should be based on connector to frame failure at these loads.

TABLE 4-4
MAXIMUM HORIZONTAL DESIGN LOAD FOR BREAKAWAY WALLS

DESIGN WIND VELOCITY	HORIZONTAL LOAD
100	15 psf
110	18 psf
120	22 psf

Figure 4-9 shows a breakaway wall detail that can be used for piles spaced 8 and 10 feet on centers. It is important to note that the only attachment to the structural frame is by nailing the end studs of the panel into the piles. In addition, it is important that number and size nails in Table 4-5 are not exceeded as the basis for the table is that the wall will fail when the load exceeds the horizontal loads indicated in the first column.

FIGURE 4-9
BREAKAWAY WALL

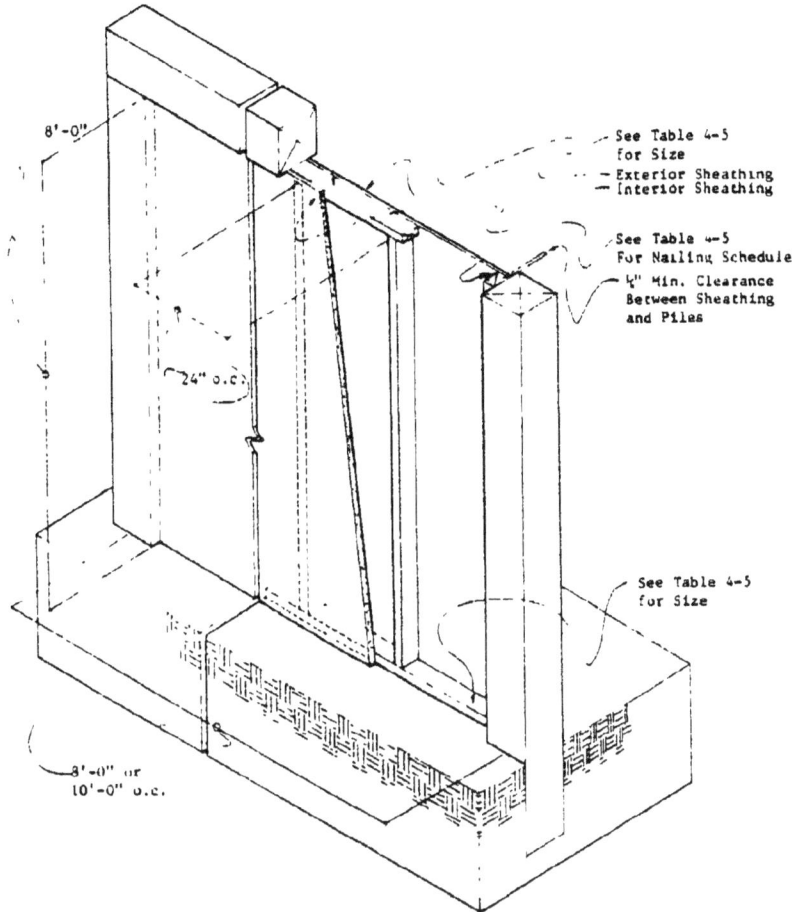

FIGURE 4-9
BREAKAWAY WALL

TABLE 4-5
PLATE AND NAIL SIZES

HORIZONTAL LOAD	PLATE SIZE PILE SPACING		NAILS – NUMBER AND SIZE PILE SPACING 3	
	8'-0"	10'-0"	8'-0"	10'-0"
15 psf	1 2 x 4	1 2 x 6	3-8d	4-8d
18 psf	1 2 x 4	1 2 x 6	4-8d	4-8d
22 psf	2 2 x 4	1 2 x 6	4-8d	5-8d

1. Stud grade or #3 surfaced dry Southern Pine or Douglas Fir
2. Construction grade surfaced dry Southern Pine or Douglas Fir
3. Nails spaced 6" from corners with intermediate nails equally spaced.

E. Stairways

Since stairways may be subject to storm surge and wave action, the stairway should be designed and constructed to breakaway at the maximum horizontal forces indicated in Table 4-4.

F. Beams

Beams support the habitable portion of the structure and transmit all loads to the foundation. In addition, in pile foundations the beams tie the tops of the piles together causing the piles to act as a structural system. These members are either solid timbers (e.g., 4x10, 4x12) or built up members using 2" wide nominal lumber (e.g., 2 - 2x10, 2 - 2x12). Splices of built up members should only occur over supports as any splicing of a member substantially reduces the strength of the beam. Table 4-6 provides minimum beam sizes based on the number of stories, beam spans and spacing velocities. Since the dead loads plus live loads exceed the wind loads for 120 mph wind velocity, the beam sizes cannot be reduced for lesser velocity winds.

TABLE 4-6
BEAM SIZES

Number Stories	Span	Spacing	Beam Size
1	8'-0"	8'-0"	3 - 2x12 [1]
2	8'-0"	8'-0"	3 - 2x12 [2]
1	10'-0"	10'-0"	3 - 2x12 [2]
2	10'-0"	10'-0"	4 - 2x14 [2]

1. Timbers to be #3 surface dry Southern Pine or Douglas Fir
2. Timbers to be #2 surface dry Southern Pine or Douglas Fir

V. BEAM CONNECTIONS TO PILES OR POSTS

In order for the beam to transmit the load to the foundation, the connection between the foundation and the beam is extremely important. This connection to piles should be made by notching the pile and. connecting the beam to the pile with bolts, bolts and gusset plates, or bolts and special connectors. See Figure 4-10 through 4-12 for examples of each type of connection. The connection of the beam to wood or masonry piers or foundation walls is equally important and should be made with steel straps embedded in concrete grout or connectors. Details of the connections are shown in Figure 4-13 through 4-15. The combination of downward dead load and live load (conventional design) exceeds the upward wind load in combination with the downward dead load and the horizontal wind loads. As a result, the beams should fully bear on the piles. The design connector loads and bolts shall comply with Table 4-7.

*TABLE 4-7
MINIMUM BOLT SPACING 3-1/2", MINIMUM EDGE DISTANCE 2"
CONTINUOUS BEAM TO PILE OR POST CONNECTION LOADS AND BOLT REQUIREMENTS

	110 mph maximum	120 mph maximum
	8'-0" Maximum Span	
one story	2 - 1/2" diam. bolts 1600 pounds	2 - 1/2" diam. bolts 1930 pounds
two story	2 - 5/8" diam. bolts 2000 pounds	2 - 3/4" diam. bolts 3860 pounds
	10'-0" Maximum Span	
one story	2 - 5/8" diam. bolts 2000 pounds	2 - 3/4" diam. bolts 2410 pounds
two story	3 - 3/4" diam. bolts 2500 pounds	3 - 7/8" diam. bolts 4820 pounds

*Table 4-7 is based on beams fully bearing on piles or post. If beams are not fully bearing, the connector loads increase to 6000 pounds for one story buildings and 9000 pounds for two story buildings. In addition, when the connectors are installed, the bolt sizes and configuration must strictly adhere to the approved plans.

FIGURE 4-10
TYPICAL CONNECTION FOR CONTINUOUS BEAM BEARING ON TOP OF SQUARE PILE

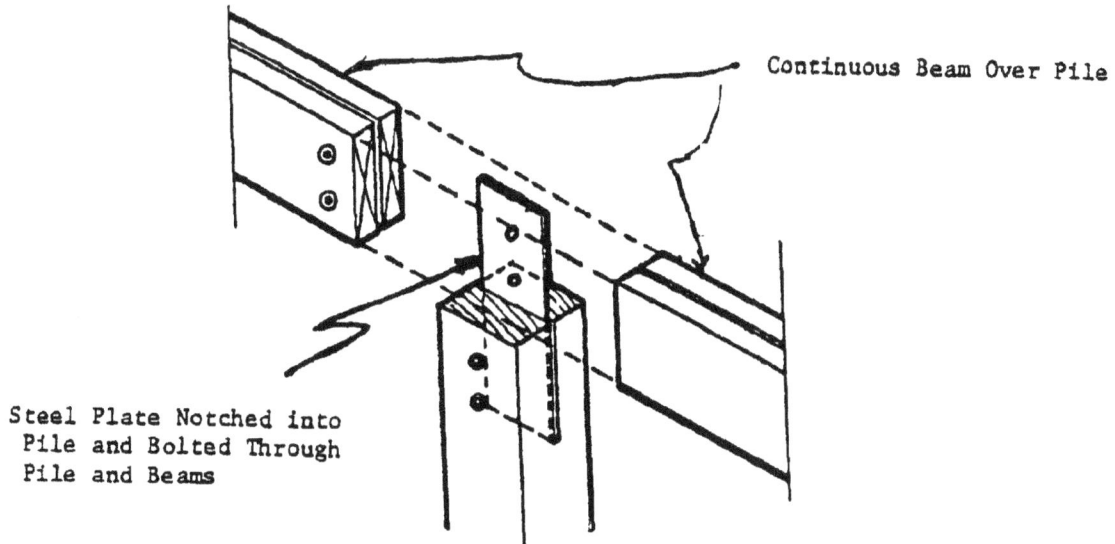

Continuous Beam Over Pile

Steel Plate Notched into
Pile and Bolted Through
Pile and Beams

FIGURE 4-11
TYPICAL CONNECTION FOR BEAM BEARING ON NOTCHED SQUARE PILE

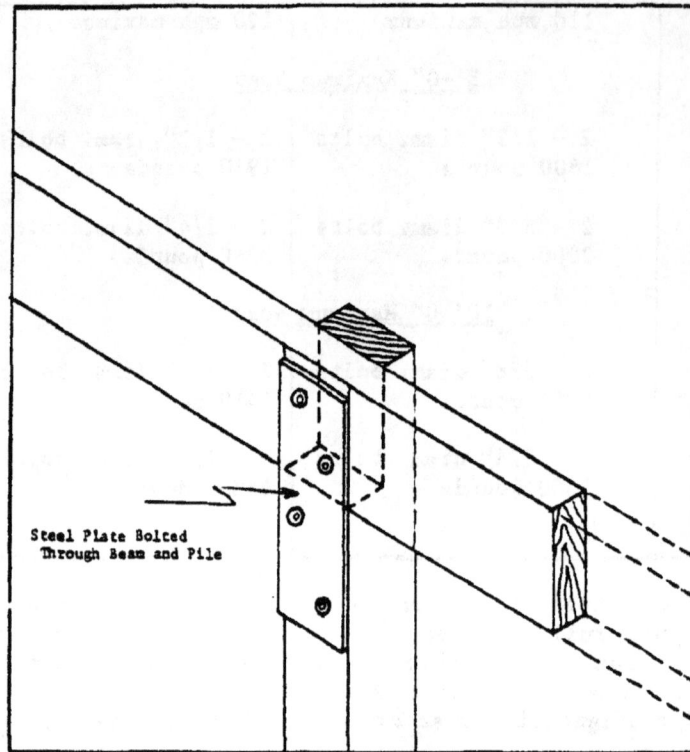

Steel Plate Bolted
Through Beam and Pile

FIGURE 4-12
TYPICAL CONNECTION CONTINUOUS BEAM ON NOTCHED ROUND PILE

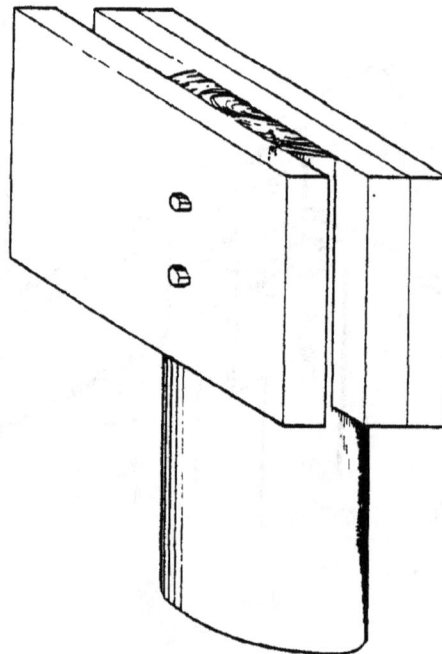

FIGURE 4-13
YPICAL CONNECTOR HOLLOW MASONRY UNIT PIER

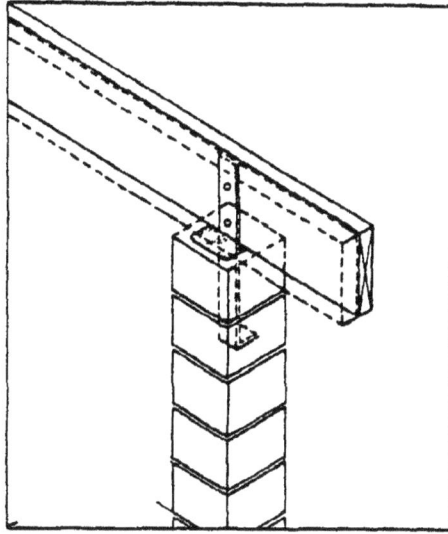

FIGURE 4-14
TYPICAL CONNECTOR FOR BRICK PIER

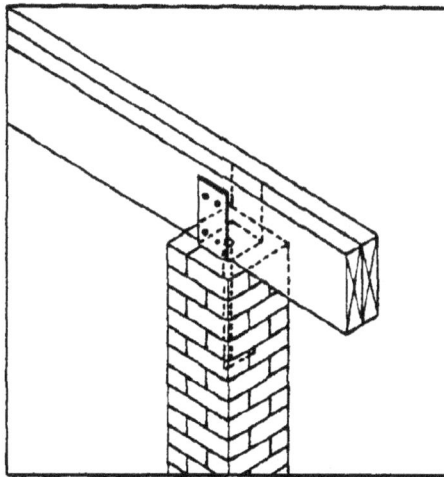

FIGURE 4-15
TYPICAL CONNECTOR WOOD POST OR PIER

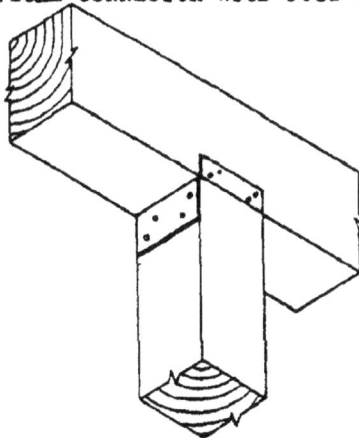

When the beam is built up and the timbers in the beam must be spliced, the splice must occur over the pile, requiring special consideration of the connection to the pile. Table 4-8 indicates the bolt sizes and spacing requirements for spliced beams. The connector loads shown in Table 4-6 are the basis for Table 4-8.

TABLE 4-8
SPLICED BEAM TO PILE OR POST BOLT REQUIREMENTS

	110 mph	120 mph
	8'-0" Maximum Span	
one story	2 - 1/2" diam. bolts each side	2 - 1/2" diam. bolts each side
two story	2 - 1/2" diam. bolts each side	2 - 1/2" diam. bolts each side
	10'-0" Maximum Span	
one story	2 - 1/2" diam. bolts each side	2 - 1/2" diam. bolts each side
two story	2 - 5/8" diam. bolts each side	2 - 5/8" diam. bolts each side

NOTES:
1. Minimum Bolts Spacing 2-1/4", Minimum Edge Distance 2"
2. Minimum End Distance 2" for 1/2" diam. bolts, 2-7/8" for 5/8" diam. bolts

It is important to note that the bolt sizes decrease in comparison to the non-spliced beam. The bolt sizes in the table should only be changed by a registered arhitect or engineer as the sizes are based on the loads in Table 4-6 and the National Design Specification for Wood Conservation. Increasing the bolt diameter would require greater end and edge distances, and spacing while decreasing the size would result in the spliced joint not supporting the design loads.

FIGURE 4-16
TYPICAL SPLICED BEAM CONNECTION

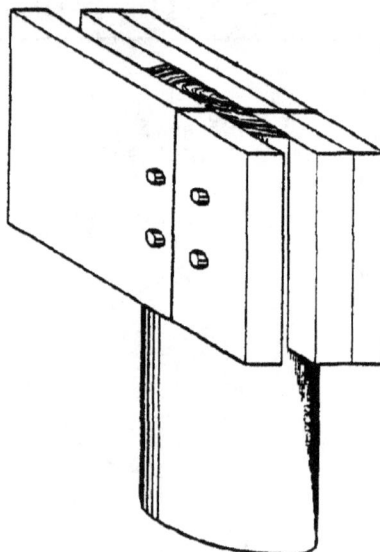

VI. FLOOR JOISTS

The primary function of floor joists is to transmit loads from the inside of the building to the beams. In elevated structures, the joists must resist the force of wind and water forces if the joists are located below the storm surge and wave height.

Wind forces may be upward or downward depending upon the configuration of the ground, and appurtenances located under the elevated structure. Water forces are upward as a result of the buoyancy forces of the water. The joists are also part of the lower diaphragm structural system that provides additional resistance to horizontal forces. Blocking should be placed between joists at all supports. This requirement is contained in most codes for conventional design. Since joists are designed for substantial live loads, no additional design requirements are recommended for coastal areas. A 2"x8" joist is adequate for spans up to 8'-0" and a 2"x10" joist is adequate for spans up to 10'-0". The joist sizes are based on a number 3 grade Southern Pine or Douglas Fir timber surface dry. These sizes are based on a maximum spacing of 24" o.c. and a 120 mph wind velocity; however, since there is a potential for reversal of loads (upward in lieu of downward), special attention must be given to the connection between the joists and the beam.

VII. JOIST TO BEAM CONNECTIONS

In conventional construction, the loads are downward. The joist is supported by a ledger strip nailed to the beam and the joist is toenailed to the ledger and beam or joist hangers. In coastal construction, the first floor joist to beam connections are subject to both upward and downward loads and should be capable of supporting both loads. See Figure 4-17 for typical joist connectors. The connectors selected for the first floor should be capable of supporting upward or downward loads for the joist spacing and spans indicated in Table 4-9. This table is applicable for design wind velocities up to 120 mph.

FIGURE 4-17
TYPICAL JOIST CONNECTORS

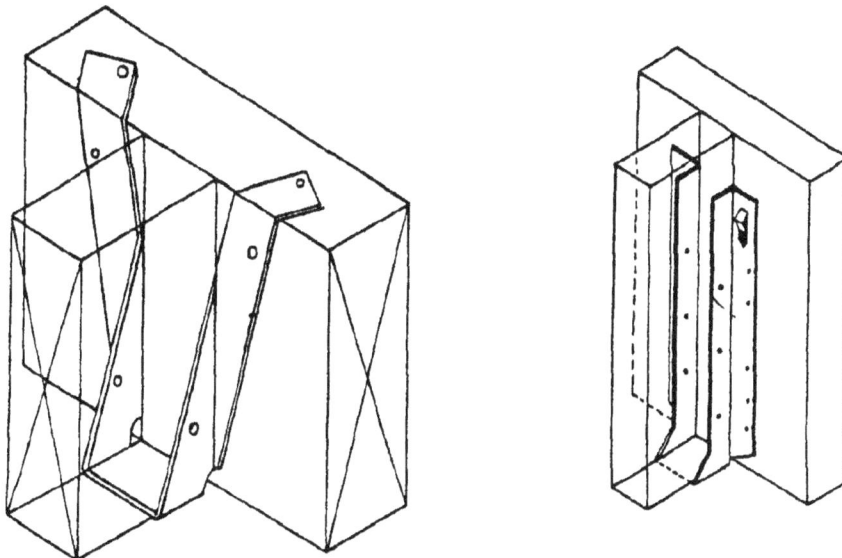

TABLE 4-9
JOIST TO BEAM CONNECTOR LOADS

Spacing	8'-0" Span	8'-0" Span 4'-0" Cantilever
16" o.c.	270 Pounds	470 Pounds
24" o.c.	400 Pounds	600 Pounds
Spacing	10'-0" Span	10'-0" Span 4'-0" Cantilever
16" o.c.	335 Pounds	535 Pounds
24" o.c.	500 Pounds	700 Pounds

Since the second floor structural system is not subject to the direct vertical forces of wind or water, conventional connections between the joists and the beam are adequate. The second floor acts as a diaphragm and supports the exterior walls, and thus, care should be taken to ensure that nailing completely conforms to conventional requirements. When joist hangers or other connectors are used, the installation of the nails or bolts must strictly adhere to the manufacturer's documents as to size and configuration.

VIII. SUBFLOORING

The primary purpose of the subflooring is to support the loads inside the structure and transfer these loads to the joists. In elevated structures, the subflooring must resist the force of wind and, in areas subject to the coastal flooding, water forces if the floor is located below the storm surge and wave height. The wind forces may be upward or downward depending upon the configuration of the grade and appurtenances located under the elevated structure. Water forces are upward from the buoyancy forces of the water. The subflooring is also a component of the floor diaphragm which plays a vital role in the resistance of horizontal forces. The conventional live load required for subflooring is substantial, so no additional loads are recommended. Conventional requirements permit the use of one inch nominal thick lumber, plywood or particleboard as subflooring. In coastal areas because of the high moisture, both the plywood and particleboard should be of the type required for exterior use. Since conventional design requirements are for downward loads regular nails are driven directly into the joists. To counteract potential reversal of these loads, annular ring or deformed shank nails should be used to attach the subflooring to the joists.

IX. WALLS

General

The primary purpose of the exterior walls in any building is to protect the occupants and contents of the buildings from the elements. In addition, the walls resist external horizontal forces such as wind, and may support the roof of the building. Walls are frequently identified by the structural elements in the wall that support or resist these forces - stud, concrete or masonry walls. In conventional construction, both metal and wood studs are used. Since concrete walls normally require extensive forming, their use in one and two family dwellings and small commercial buildings is normally limited to basement and foundation walls. Masonry walls may be nonreinforced or reinforced and composed of hollow masonry units of sand or man-made light weight aggregate or solid brick masonry units.

A. Metal Studs

In coastal construction, metal studs are rarely used because of their corrosive nature and lack of industry standards covering structural characteristics. In the event steel studs are used, they should be heavily galvanized, and the structure designed by a registered engineer or architect.

B. Wood Studs

Wood studs are the most frequently used structural elements for one and two family dwellings in coastal areas. The light weight of a wood stud wall makes it the most practical wall for one and two family dwellings constructed on piles. A 2x4 stud or number 3 grade Southern Pine of Douglas Fir timber surface dry 16" o.c. is adequate for nominal 8'-0" high walls in one and two story buildings for wind velocities not exceeding 120 mph. Section 105 of Appendix A of this manual provides requirements for different stud spacing.

C. Concrete Walls

High ground water tables often prohibit basements in coastal areas. Thus, concrete foundation walls are only used in one and two family dwellings and small commercial buildings under unusual conditions. If concrete walls are required, they should be designed by a registered engineer or architect.

D. Masonry Walls

Masonry walls are heavy when compared to stud construction and subject to cracking when the structural system supporting the wall deflects. As a result masonry walls should only be used where continuous spread footings are used. In coastal areas, all masonry walls should be reinforced. If the wall is constructed of masonry units, the minimum width of the wall should be 8 inches nominal for hollow units and 6 inches for solid units. See Figure 4-18 for bond beam and reinforcement requirements for hollow masonry unit walls walls to be constructed in areas with a maximum design wind velocity of 120 mph.

FIGURE 4-18
HOLLOW MASONRY WALL

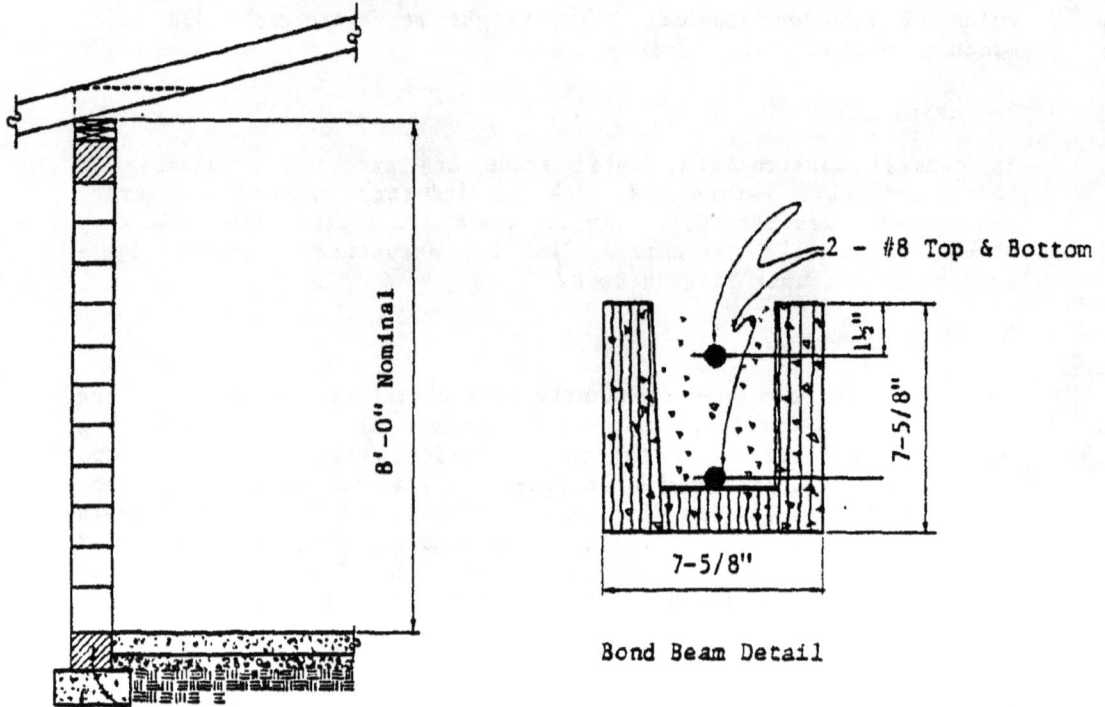

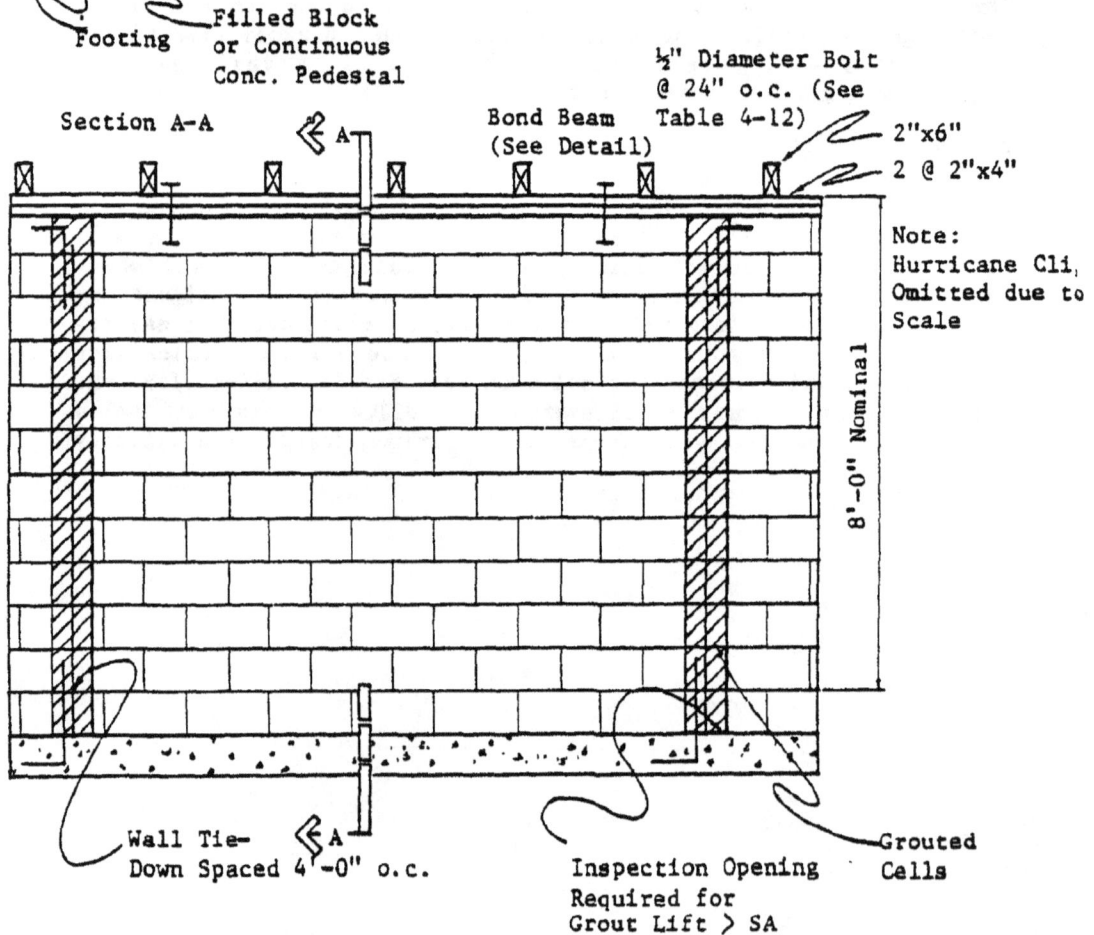

8'-0" Nominal

Footing

Filled Block
or Continuous
Conc. Pedestal

Section A-A

2 - #8 Top & Bottom

1½"

7-5/8"

7-5/8"

Bond Beam Detail

A

Bond Beam
(See Detail)

½" Diameter Bolt
@ 24" o.c. (See
Table 4-12)

2"x6"

2 @ 2"x4"

Note:
Hurricane Cli.
Omitted due to
Scale

8'-0" Nominal

Wall Tie-
Down Spaced 4'-0" o.c.

A

Inspection Opening
Required for
Grout Lift > SA

Grouted
Cells

Figure 4-19 provides bond beam and reinforcement requirements for solid masonry walls constructed in areas with a maximum design wind velocity of 120 mph. The top plate size and connection requirements to the bond beams are provided in Table 4-12.

FIGURE 4-19
SOLID MASONRY WALL

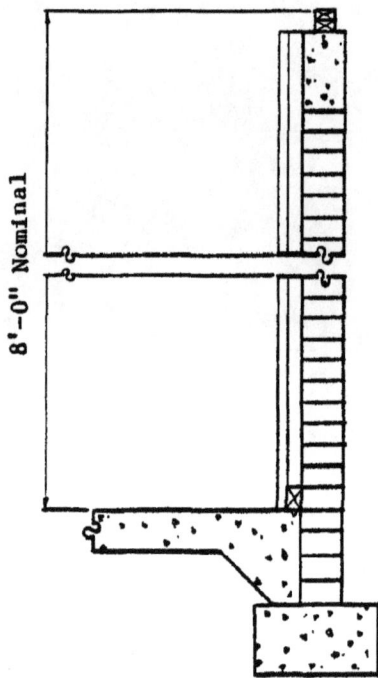

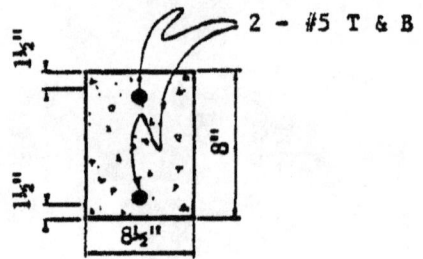

Section A-A

8'-0" Nominal

2 - #5 T & B

1½"   8"   1½"   8½"

A
Bond Beam Detail

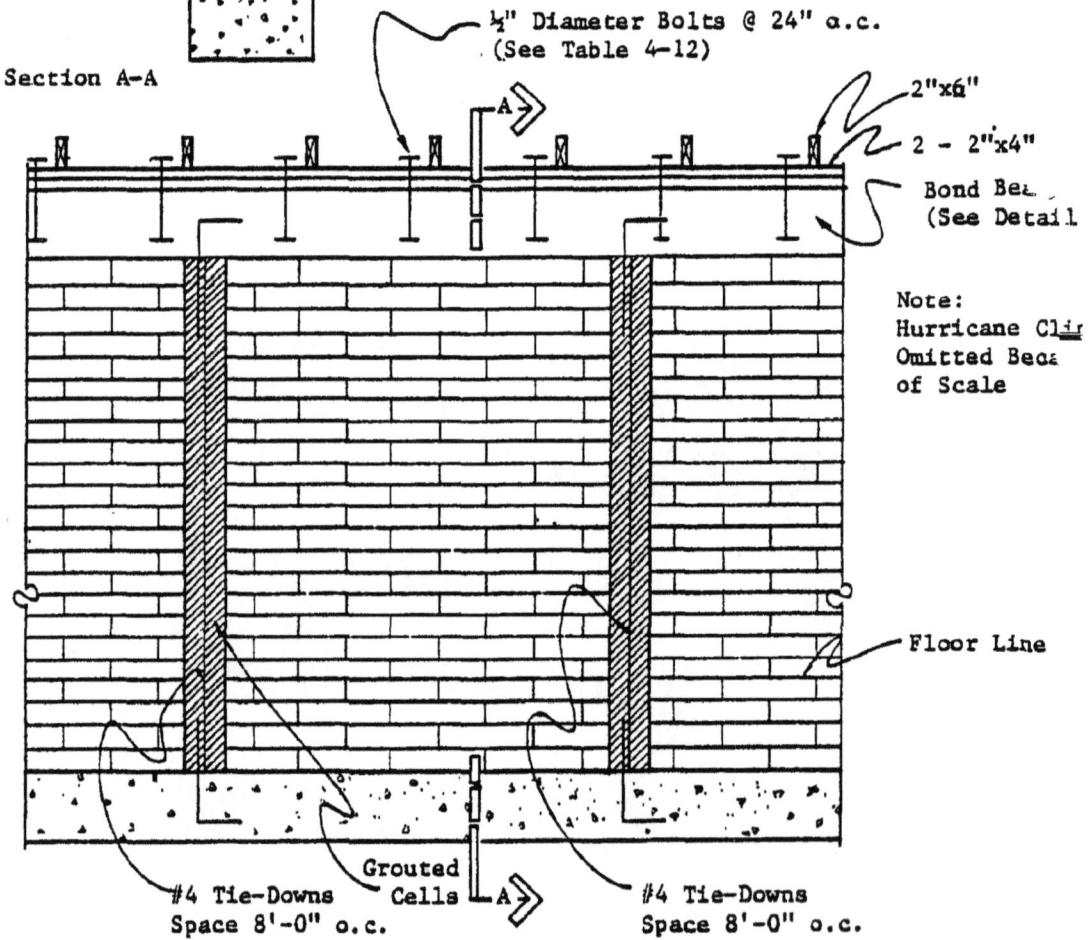

½" Diameter Bolts @ 24" o.c.
(See Table 4-12)

2"x6"

2 - 2"x4"

Bond Beam
(See Detail

Note:
Hurricane Clip
Omitted Because
of Scale

Floor Line

#4 Tie-Downs
Space 8'-0" o.c.

Grouted
Cells

#4 Tie-Downs
Space 8'-0" o.c.

Sections 106 and 107 in Appendix A of this manual provide bond beam and reinforcement requirements for design wind velocities other than 120 mph.

The mortar for masonry walls in coastal construction should be either type M or S and conform to the requirements of Table 4-10.

TABLE 4-10
MORTAR PROPORTIONS BY VOLUME

Mortar Type	Portland Cement Cu.Ft.	Masonry Cement Cu.Ft.	Hydrated Lime or Lime Putty Cu.Ft.	Aggregate Measured In Damp Loose Condition Cu.Ft.
M	1	None	1/4	Not less than 2-1/4 and not more than 3 times the sum of the volumes of cement and lime used.
M	1	1	None	
S	1	None	Over 1/4 to 1/2	
S	1/2	1	None	

Mortar should be mixed in a mechanical batch mixer for 5 minutes, with the amount of water required to produce the desired workability.

Grout for hollow masonry construction should be proportioned within the limits of Table 4-11.

TABLE 4-11
GROUT PROPERTIES BY VOLUME

Mortar Type	Portland Cement Cu.Ft.	Hydrated Lime or Lime Putty Cu.Ft.	Aggregate Measured in Damp Loose Condition	
			Course	Fine
Fine Grout	1	0 to 1/10		2-1/4 to 3 times the sum of the volumes of cement and lime used.
Coarse Grout	1	0 to 1/10	1 to 2 times the sum of the volumes of cement and lime used.	

The amount of water added should be sufficient to bring the mixture to the desired consistency.

E. Wall-Footing or Beam Connection

As previously discussed, the loads acting on the walls must be transferred to the footings, or in pile construction, the supporting beams between the piles. In both solid and hollow masonry construction, this connection is accomplished by making a standard 90° hook in the end of the vertical walls reinforcement and extending the hook 6 inches into the footing (Figures 4-18 and 4-19). In wood stud construction, the transfer of the load from the wall to the foundation is accomplished by connecting the stud to a bottom plate and connecting the plate to a footing, foundation wall beam or floor joist. This bottom plate foundation walls and footings connection requirements for masonry and concrete construction are shown in Table 4-12.

TABLE 4-12
TOP AND BOTTOM PLATE BEAM CONNECTION REQUIREMENTS
IN MASONRY AND CONCRETE

Velocity MPH	Plate Size	Wood Grade and Species	Bolt Size [2]	Bolt Spacing [2]	Washer
110	2-2"x4"	1	1/2"	24" oc	3"x3"x1/4" Plate
120	2-2"x4"	1	1/2"	24" oc	3"x3"x5/16" Plate

NOTES:
1. Minimum grade-stud grade Southern Pine or Douglas Fir Surface Dry
2. ASTM A 307 Bolts with standard 90° hook and embedded in concrete a minimum of 6 inches.

When the bottom plate is connected to platform construction, the connection must be made with connectors from the bottom plate to the floor joists or rim joists. In most cases, it is advantageous to use a single connector that is designed to jointly connect the stud, plate and joist. A detail of this connector is shown in Figure 4-20. If the connectors are placed at the joint between each stud and the plate, a single 2"x4" plate may be used. The connector selected shall be designed to resist the upward force shown in Table 4-13. When the connector is installed, the nail sizes and configuration must strictly adhere to the manufacturer's documentation, as this is the point at which the vertical forces acting on the roof are transferred to the floor joist. Surveys have indicated that this connection is frequently omitted, resulting in the living area being totally blown off the platform.

FIGURE 4-20
EXAMPLE OF TYPICAL RAFTER TO PLATE TO STUD CONNECTOR

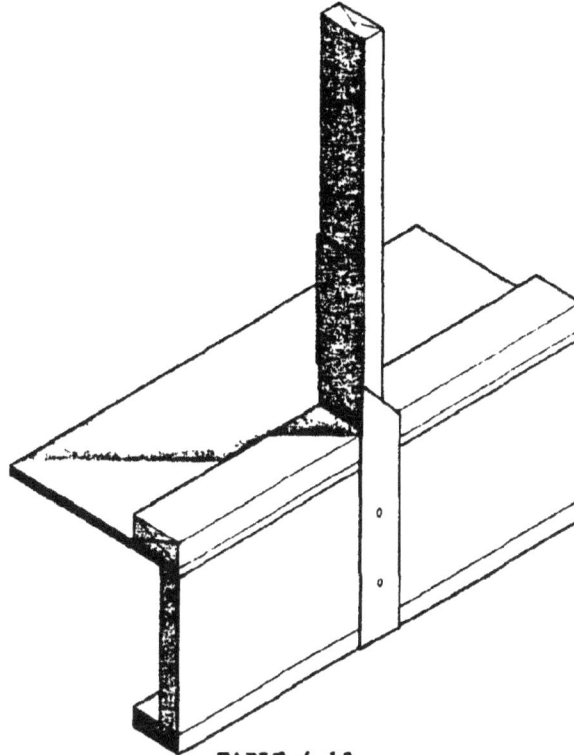

TABLE 4-13
CONNECTOR LOADS (LBS) FOR
VARIOUS RAFTER SPANS
AND OVERHANGS

BUILDING WIDTH	110 MPH VELOCITY OVERHANG			120 MPH VELOCITY OVERHANG		
	0'	2'	4'	0'	2'	4'
20'	215	285	360	260	345	435
24'	260	330	400	310	400	485
30'	320	395	465	390	475	565
32'	345	415	485	415	500	585
40'	430	500	575	520	605	690

*Loads based on 16" oc rafter spacing, 9 psf dead load and rafter slope between 0° - 40°. For 24" spacing, increase loads by 50%.

F. Two Story Construction

In two story construction, it is essential that the second floor studs, sole plate, and joists are connected to the exterior wall or interior partition plates and studs, as this is a point in the building that transfers the vertical roof loads to the floor framing system and foundation. This is best accomplished by a single connector that is designed to connect all of the elements. The connector selected should be designed to resist the upward or downward force shown in Table 4-13. See Figure 4-21 for typical detail of a single connector.

29-4

FIGURE 4-21
EXAMPLE OF TYPICAL CONNECTOR FOR TWO STORY BUILDING

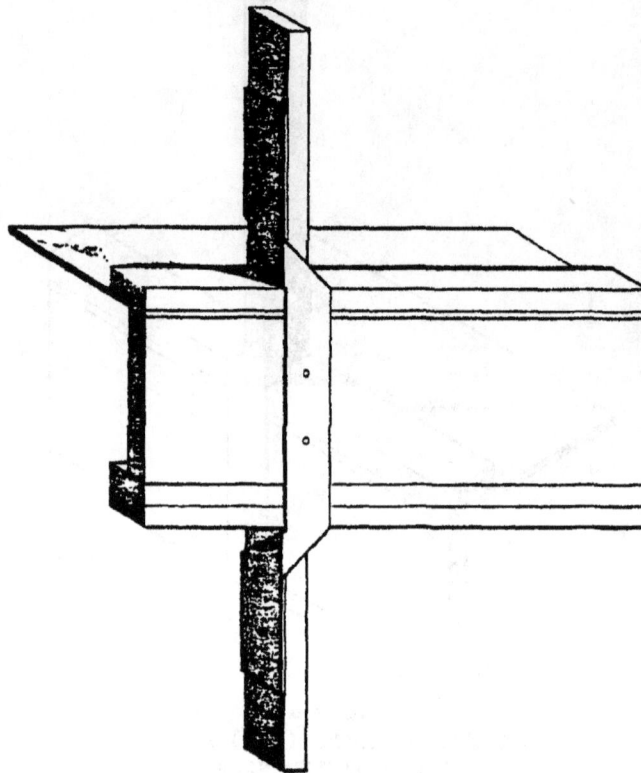

G. Wall Sheathing

Sheathing is used in wood or metal stud construction to transmit the
horizontal forces acting on the exterior covering to the studs.
Normal materials such as wood, plywood, particleboard and insulated
fiber materials are used as sheathing. Sheathing must develop the
necessary shear resistance to allow exterior walls to support the
horizontal loads. The sheathing should be 15/32" plywood or
1-1/2" particleboard nailed in accordance with Table 4-14. The
plywood or particleboard is to extend a minimum of 4'-0" from each
corner. Fasteners for conventional sheathing such as plywood and
particleboard are designed for inward or lateral loads. Since wind
may produce either inward or outward loads, deformed shank
or annular nails should be used to attach the sheathing in coastal
areas.

TABLE 4-14
NAIL SIZE AND SPACING
FOR CORNER SHEATHING

| | 110 mph maximum | | | | 120 mph maximum | | | |
| | Particleboard [1] | | Plywood [2] | | Particleboard [1] | | Plywood [2] | |
Width	Size	Spacing	Size	Spacing	Size	Spacing	Size	Spacing
20	10d	6" oc	10d	6" oc	10d	6" oc	10d	6" oc
24	10d	6" oc	10d	6" oc	10d	6" oc	10d	6" oc
30	10d	6" oc	10d	6" oc	10d	6" oc	10d	6" oc
32	10d	6" oc	10d	6" oc	10d	6" oc	10d	6" oc
40	10d	4" oc	10d	4" oc	10d	4" oc	10d	4" oc

NOTES:
1. Minimum Grade particleboard - 1/2" - 2M-W
2. Plywood Grade - 15/32" C-D Sheathing, C-C Exterior and Panel Siding

H. Exterior Wall Coverings

Exterior wall coverings include many materials. Conventional design requirements require only consideration of inward loadings. However, since wind loads may produce both inward and outward forces plus missiles, special care should be observed in coastal construction when selecting both the exterior wall material and the connection of the covering to the sheathing and/or studs. The conventional requirements for fasteners are adequate for the inward wind loads, but if nails are used, they should be either deformed shank or annular nails to resist the outward wind forces. The manufacturer of wall panels should provide information on the wind forces the panel will withstand, depending on spacing of supports. The manufacturer's data should be based on ASTM Standard E 330, Test for "Structural Performance of Exterior Windows, Curtain Walls and Doors by Uniform Static Air Pressure Difference".

Conventional wall covering materials with the exception of glass and some plastics have the strength characteristics to resist the impact of missiles. Several tests have been proposed to determine the resistance to impact of exterior wall materials - one for example consists of dropping a 2x4 timber weighing 40 pounds on a test specimen. The test specimen of cladding or glazing material would receive two impacts: one at the center of the specimen and the other at a corner 6 inches from each edge of the specimen measured from points free of any supporting members. Acceptable materials would reject the timber without penetration. Since none of the tests developed to this point are refined to the point that repeatable results can be expected, the recommendation is that nonbrittle materials should be used for exterior wall coverings.

Masonry veneer is considered another exterior wall covering. The veneer should not be attached to wood at any point more than 25 feet above the foundation. It should be attached to the studs with hot dipped corrosion-resistant metal ties. The ties should not be less than Number 6 U.S. gauge wire and should have a hook embedded in the

mortar joint, or of sheet metal not less than 22 U.S. gauge
corrugated. Each tie should be spaced not more than 24 inches on
center horizontally and support not more than 2 square feet of wall
surface. Masonry veneer may be supported only by noncombustible
members.

I. Windows and Doors

The failure of windows and doors has been one of the most frequent
building component damages observed after a hurricane has passed
over an area. These components fail in two ways, (1) breaking of
the glazing by either wind or water borne debris, and (2) the
failure of the door or window frame, hinges and locks to support the
horizontal forces of the wind on the door or window.

The most frequently used method of protecting glass in windows or
doors is the installation of shutters designed to resist impact
forces. It is recommended that these shutters be constructed of
minimum 1" nominal lumber, or 15/32" plywood or equivalent
materials.

The most practical way to eliminate door and window failure from
horizontal wind forces is the selection of only those doors and
windows where the manufacturer provides data that indicates the
specific wind forces for which that door or window has been designed
and tested, and furnishes instructions for the installation of the
door or window. The builder, contractor and inspector should ensure
that these components are installed in accordance with these
instructions.

X.   ROOF

The purpose of the roof is to protect the occupants from the
elements of the environment, transfer forces acting on the roof to
the walls and provide the top wall support. The roof consists of
rafters, trussed rafters or trusses, sheathing and exterior
covering.

Many rafters, trussed rafters and some trusses in conventional
construction are designed only for vertical downward loads. Except
in coastal areas, the vertical downward design dead and live load
normally exceeds the vertical upward design wind loads. Also,
in sloped roofs, the ceiling joists and the bottom chord of the
truss resist the horizontal roof force at the top of the wall. This
joist or chord is also needed to support the ceiling. In the event
of an internal pressure, the joist or chord would resist the upward
forces created by the internal pressure.

Table 4-15 indicates rafter size for coastal areas where the design
wind velocity does not exceed 120 mph. The table is based on
surface dry Douglas Fir or Southern Pine timbers.

32-4

TABLE 4-15
RAFTER SIZE BASED ON SPAN @ 16" o.c.

Rafter Size	Maximum Span #3 Grade	Maximum Span #2 Grade
2 x 6	10'	12'
2 x 8	12'	16'
2 x 10	16'	20'
2 x 12	20'	24'

In addition to timber size, the connections of truss joints are critical in truss design and selection. When trusses are used, the producer or manufacturer should supply the design requirements for each specific truss rafter or truss type used in a building. The contractor, builder and inspector should review the data to ensure that the design is adequate for the wind (uplift) loads required for the area in which the building is to be constructed.

Conventional ceiling joints are used to support the load that may be imposed as the result of storing items in the attic space and resist horizontal loads at the top of walls. Normally the downward design live load is 10 psf or 20 psf, which in both cases would exceed the upward internal wind design pressure in a building. Based on this, the sizes of conventionally designed ceiling joists are adequate for coastal areas. At the gable ends the ceiling acts as a diaphragm and supports the exterior wall at the top. Analysis of wind failures have indicated the buckling of the ceiling materials has occurred in some buildings, leading to failure of the gable end wall. It is recommended that 2"x4" blocking spaced 2'-0" on centers be placed between the joist for 8'-0" from each gable end.

A. Connectors

The rafter, truss rafter and truss connection to the wall is critical in hurricane resistant design. If this connection is inadequate, there is a great potential for the loss of the roof and subsequent collapse of the exterior walls. The roof structure may rest directly on a bond beam or, most frequently, on a wood top plate. In either case, the connector should be able to resist an upward or downward force shown in Table 4-13. The sizes of the wood plate, bolts, washers, and spacing (See Table 4-12). When the connectors are installed, the size and configuration of the bolts and nails must strictly comply with the manufacturer's documentation. A typical connector is shown in Figure 4-22. Figure 4-23 is an example of connector frequently used in coastal areas. If this connector is used, another connector must be used to connect the upper and lower top plates together, and both of these top plates to the studs.

FIGURE 4-22
EXAMPLE OF TYPICAL RAFTER TO WALL CONNECTION

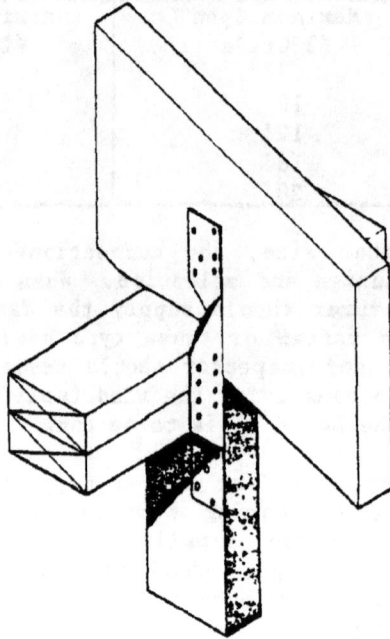

FIGURE 4-23
EXAMPLE OF RAFTER TO UPPER TOP PLATE CONNECTOR

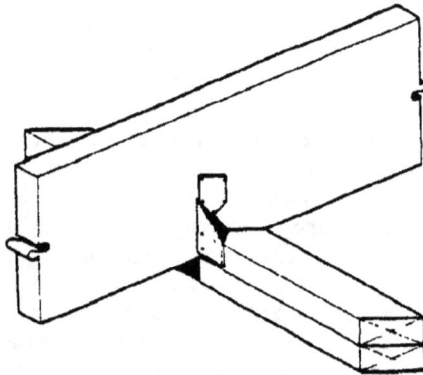

B. Roof Sheathing

The purpose of the roof sheathing is to provide vertical support for roof live loads and transmit these forces to the rafters, trussed rafters or trusses. In addition, the sheathing provides lateral resistance for the top of the roof framing system. In conventional construction, the sheathing and connections are normally designed for vertically downward loads. The conventional downward loads are adequate for coastal areas. However, wind forces will cause upward forces, and thus deformed shank or annular nails should be used to attach the roof sheathing. In several instances, analyses of wind failures have indicated that solid sheathing of plywood less than 15/32" thick or particleboard less than 1/2 inch thick buckles under horizontal wind forces on the gable wall. See Figure 4-19. It is recommended that a minimum of 15/32" thick plywood or 1/2" thick particleboard sheathing or 2x4 blocking spaced 2'-0" on center be placed between the rafters for the first 8 feet from each roof gable end. Spaced sheathing is normally a minimum of 1 inch nominal thickness, however, because of the distance between the spaced sheathing the sheathing is subject to buckling at the gable ends; therefore, it is recommended that 2x4 blocking spaced 2'-0" on center be placed between the rafters for the first 8 feet from the roof gable ends.

FIGURE 4-24
ROOF SHEATHING FAILURE

C. Roof Coverings

The primary purpose of a roof covering is to provide weather protection at the roof. Since the roof sheathing in conventional design transmits the vertical downward load to the roof structural system, manufacturers in general only indicate design loads for roof coverings that are installed without roof sheathing. These design loads are vertical downward and exceed wind load requirements. In most cases the wind loads on a roof in coastal areas are vertically upward. Thus, when nails are used to fasten the covering, deformed shank or annular nails should be used.

When the roof coverings are applied over sheathing in coastal areas, the manufacturer should be required to furnish data to indicate the roof covering to be used has been designed and tested to resist the designed wind forces. The term wind resistance as it applies to composition shingles (e.g., asphalt fiberglass, etc.) indicates that the shingles have been tested for wind in accordance with U.L. Standard 997, "Wind Resistance of Prepared Roof Covering Materials". This standard requires testing to a 60 mph wind over a two hour time period. Although the two hour time period would normally exceed the time period a building would incur maximum hurricane winds, the 60 mph requirement is not adequate for hurricane requirements and is not an acceptable test for coastal areas.

XI. DECKS AND PORCHES

Decks and porches should be designed and constructed with the applicable section of this chapter. Typical examples would be that deck piles should comply with Section 2D, Beams, Section 4F, Beam Connections, Section V, etc. Analysis of porch failures indicate that most have occurred as the result of not properly connecting the porch roof to the post or the post to the foundation. The connector loads shown in Table 4-7 should be used to size connections.

XII. MECHANICAL EQUIPMENT

Mechanical equipment for heating and cooling is frequently located on concrete ground slabs under the building or adjacent to the building. In addition, cooling condensers are frequently located on platforms at the first floor level. This elevated platform location minimizes the potential of the equipment being damaged from storm surge and wave action. In a number of cases, these platforms were found not to be designed or constructed to resist hurricane wind forces. In fact, the winds from Hurricane Alicia blew a number of these platforms away.

It is recommended that the elevated platforms be used for areas subject to storm surge and the platform be constructed to resist hurricane force winds.

Figure 4-25 indicates the minimum lap length, nail size and spacing for wind loads up to 120 mph. These dimensions and sizes are based on a maximum platform width of 4'-0". A 2"x8" number 3 grade surface dry Southern Pine or Douglas Fir Joist is adequate for 8'-0" spans, while a 2"x10" is required for 10'-0" spans. These sizes are based on a maximum spacing of 24" oc, a maximum cantilever span of 4'-0" and a maximum wind velocity of 120 mph.

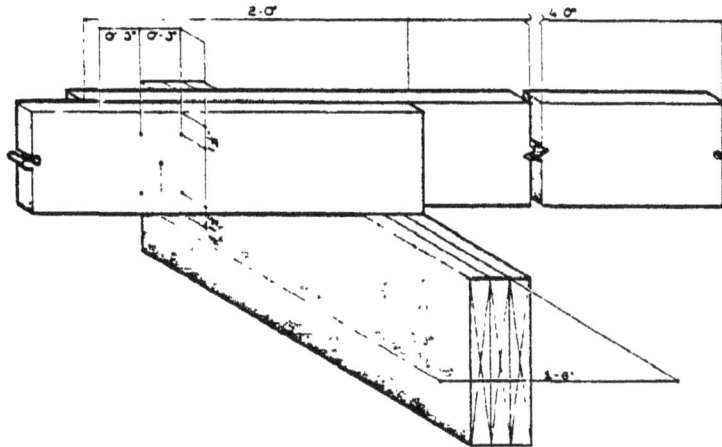

FIGURE 4-25
CANTILEVERED EQUIPMENT PLATFORM

XIII. WATER, SEWAGE AND ELECTRICAL SERVICES

Water and sewage services are provided from the ground below the living area. In order to ensure the minimum impact on these services from storm surge and wave actions, the plumbing.providing these services should be installed in contact with and supported by the piling. Electrical service may be provided overhead or underground. If the service is underground, the service should also be installed in contact with and supported by the piling. If the service is overhead, the lines and connectors should be designed in accordance with the wind load requirements for the area. The local utility service should be contacted for these design requirements.

XIV. DESIGN EXAMPLE

This Chapter provides tables and other prescriptive information that when properly used, will minimize the damage from a hurricane. To best understand the use of this information a single family residence is to be designed for Galveston, Texas. The site and floor plan are shown on Figures 4-26 and 4-27 and are typical of a one story residential building being constructed along the U. S. coasts. This example will follow in the order the various structural elements presented in Sections II through XII of this Chapter.

FIGURE 4-26
SITE PLAN

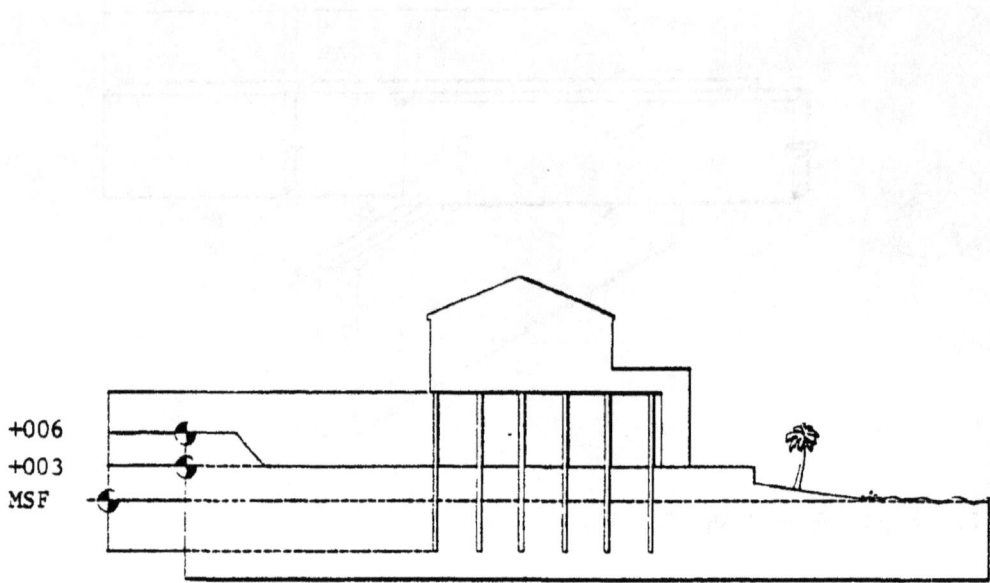

+006
+003
MSF

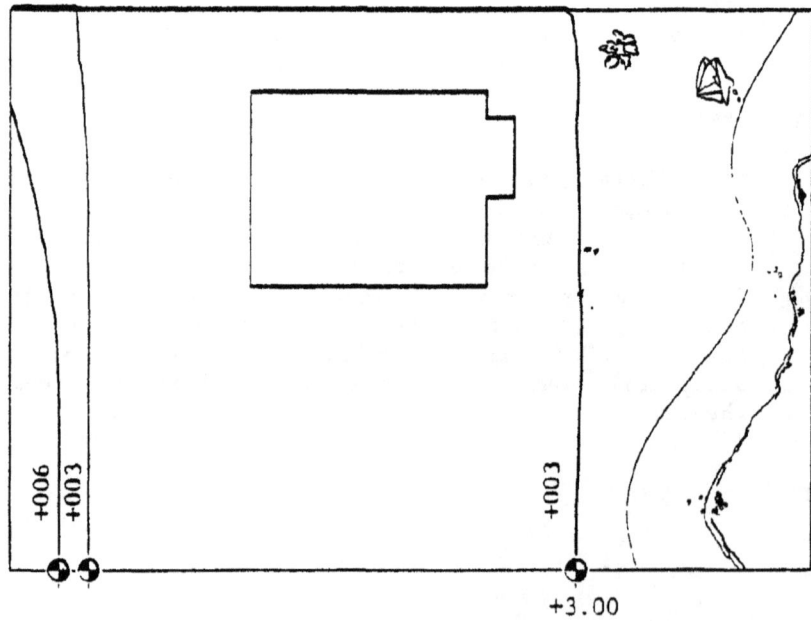

+006
+003

+003

+3.00

FIGURE 4-27
FLOOR PLAN

DECK

35'-0"

6'-4" 7'-8" 6'-8" 8'-0" 6'-4"

12'-0"

6'-6" SGD   2'-9"-6"   6'-6" SGD

REF

①

KITCH.

LIVING/DINING

9'-4"

13'-4"

3'-3"

3'-4"   6'-0"   3'-4"   4'-6"   3'-4"   14'-6"

2x8 CLG JSTS @ 16"c

3'-6"

3'-0"

13'-4"

16'-4"

③

⑤

WASH

WH

⑤

⑤

3'-0"

3'-8"

⑤   UTIL.   DRY   ⑤   ⑤   HALL   ③

32'-0"

③   ⑤

BATH

22'-8"

6'-4"

CLOS.

6'-8"

BR#1

⑤

BR#2

③

CLOS.

15'-8"

12'-0"

⑦   CLOS.   CLOS.   ⑦

5'-8"   5'-8"

4'-5"   3'-2"   4'-5"

4'-0"   13'-6"   13'-6"   4'-0"

12'-4"   2'-4" 5'-8" 2'-10"   11'-10"

35'-0"

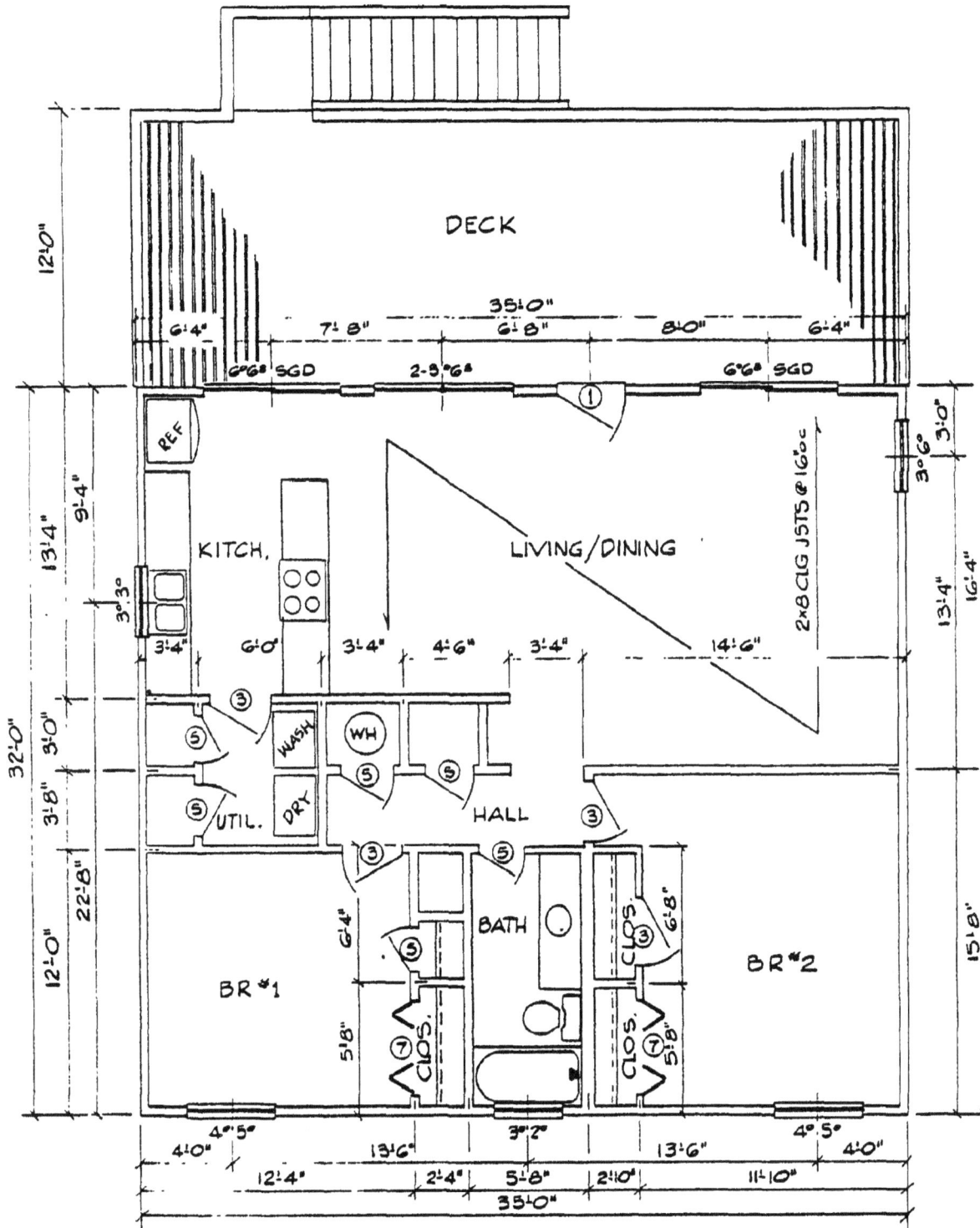

FIRST FLOOR PLAN
SCALE ~ 1/4" = 1'-0"

The first step is to determine the design wind velocity for the City
of Galveston. This may be obtained from the wind speed map in the
Standard Building Code, Figure 4-28. From Figure 4-28 a design wind
velocity of 110 mph is obtained.

FIGURE 4-28
BASIC WIND SPEED IN MILES PER HOUR

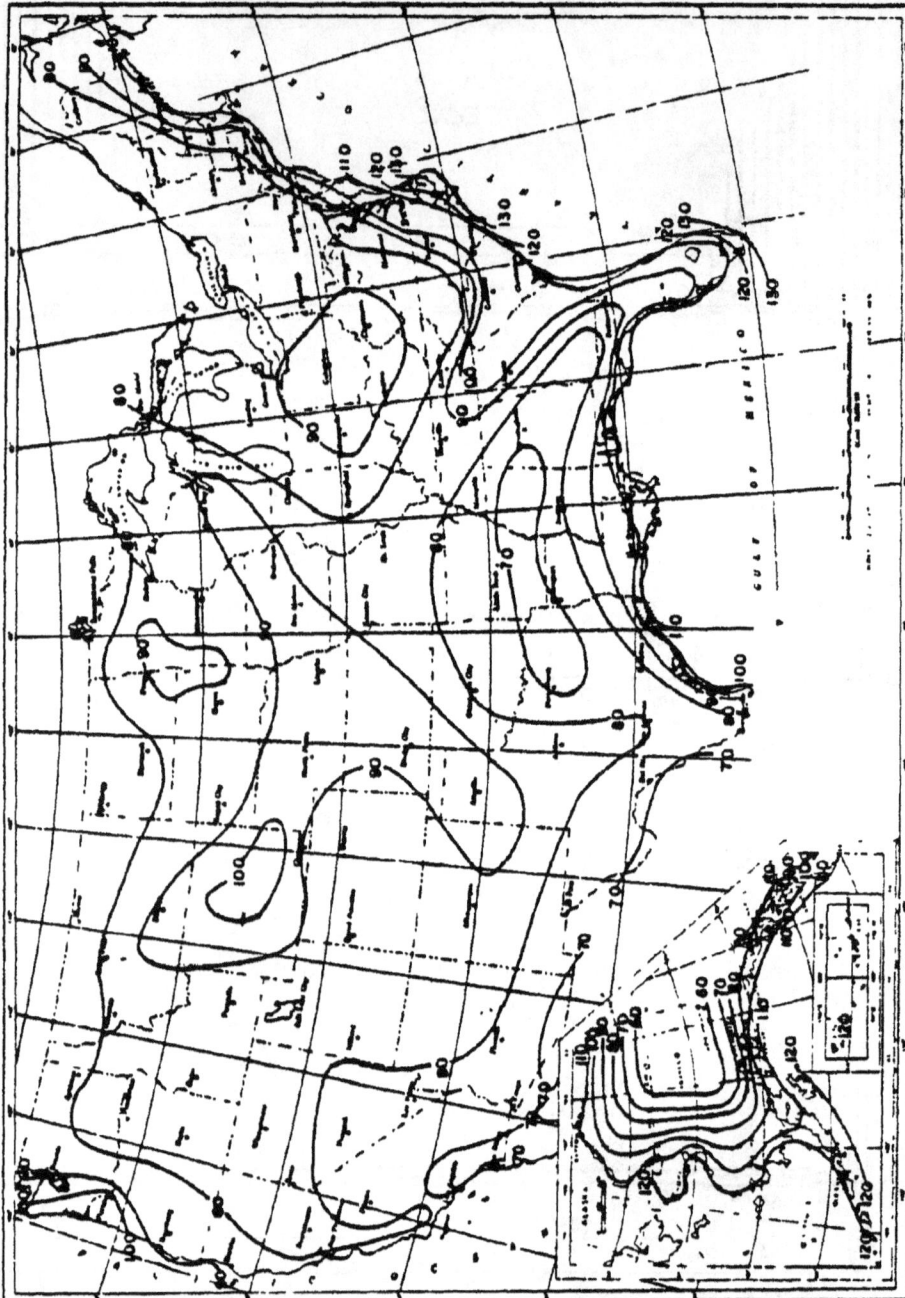

Annual Extreme Fastest-Mile Speed 30 ft Above Ground, 100 Year Mean
Recurrence Interval

NOTE: The Virgin Islands and Puerto Rico shall use a basic wind speed
of 110 mph.

Standard Building Code/1985

40-4

Since the Standard Building Code is a model code, local governments may revise the model code. Galveston is an excellent example of this as the city requires a design wind velocity of 120 mph for all structures not located behind the seawall. For the purpose of this example, we will use the wind velocities shown in Figure 4-28.

The second step is to determine the location of the building in regard to the shore line. In this case, we find this residence will be constructed in an area subject to storm surge and wave action. Using this information and the 110 mph wind velocity, we can refer to Table 4-2, Figure 4-29 and determine the minimum pile diameter to be 10 inches at the tip. From the notes in table one, we determine that the first floor elevation may be a maximum 16 feet above the mean sea level and the pile must penetrate 12 feet below mean sea level. If the residence had been located in an area not subject to storm surge and wave action, the pile diameter could have been reduced to 8 inches with the first floor elevation being a maximum 8 feet above grade and the pile penetration 12 feet below grade. In addition, we find the pile must be Southern Pine or Douglas Fir, that the maximum spacing is 8 feet on center with a minimum of four piles in any row.

FIGURE 4-29

TABLE 4-2[2,3,4,5]
MINIMUM PILE DIAMETERS FOR SANDY SOIL

	110 mph maximum		120 mph maximum	
	Subject to Surge and Wave [1]	Not Subject to Surge and Wave [2]	Subject to Surge and Wave [1]	Not Subject to Surge and Wave [2]
	8'-0" Maximum Span			
one story	10" diam. tip	8" diam. tip	10" diam. tip	8" diam. tip
two story	10" diam. tip	8" diam. tip	12" diam. tip	10" diam. tip
	10'-0" Maximum Span			
one story	Not applicable	8" diam. tip	Not applicable	8" diam. tip
two story	Not applicable	10" diam. tip	Not applicable	10" diam. tip

Notes:

1. Minimum pile penetration 12 feet below MSL
   Maximum first floor elevation 16 feet above MSL
2. Minimum pile penetration 12 feet below grade
   Maximum first floor elevation 8 feet above grade
3. Pile size based on Southern Pine pile or Douglas Fir
4. Southern Pine or Douglas Fir square piles having the least dimension equal to the pile diameters shown in Table 4-2 may be substituted for round piles.
5. Minimum number piles in a row - 4.

Figures 4-30 and 4-31 indicate the proposed structural framing for the residence. Figure 4-32 shows the elevations.

FIGURE 4-30
STRUCTURAL DETAILS

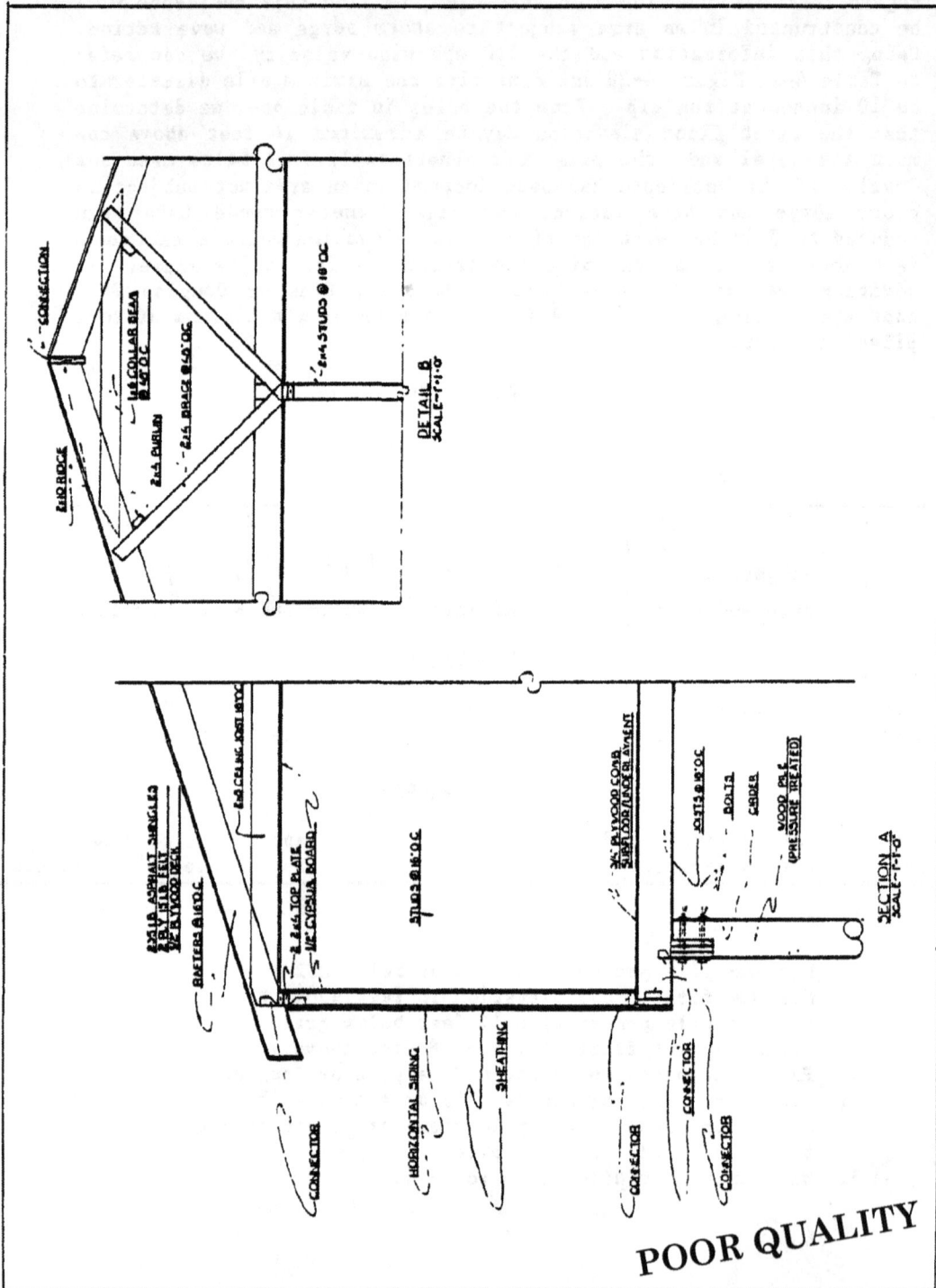

POOR QUALITY

FIGURE 4-31
STRUCTURAL FRAMING PLAN

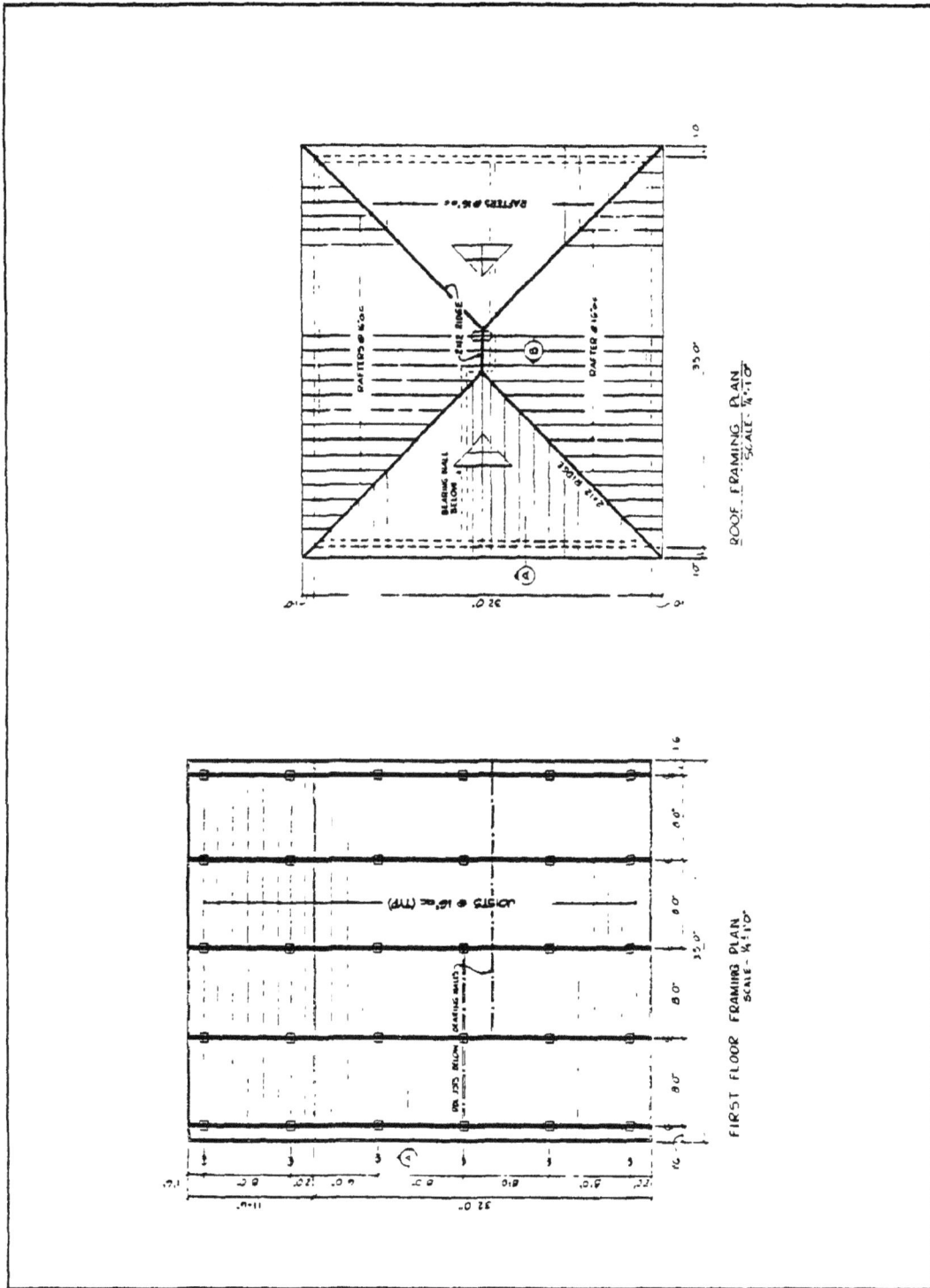

ROOF FRAMING PLAN
SCALE - 1/4" = 1'-0"

FIRST FLOOR FRAMING PLAN
SCALE - 1/4" = 1'-0"

FIGURE 4-32
ELEVATIONS

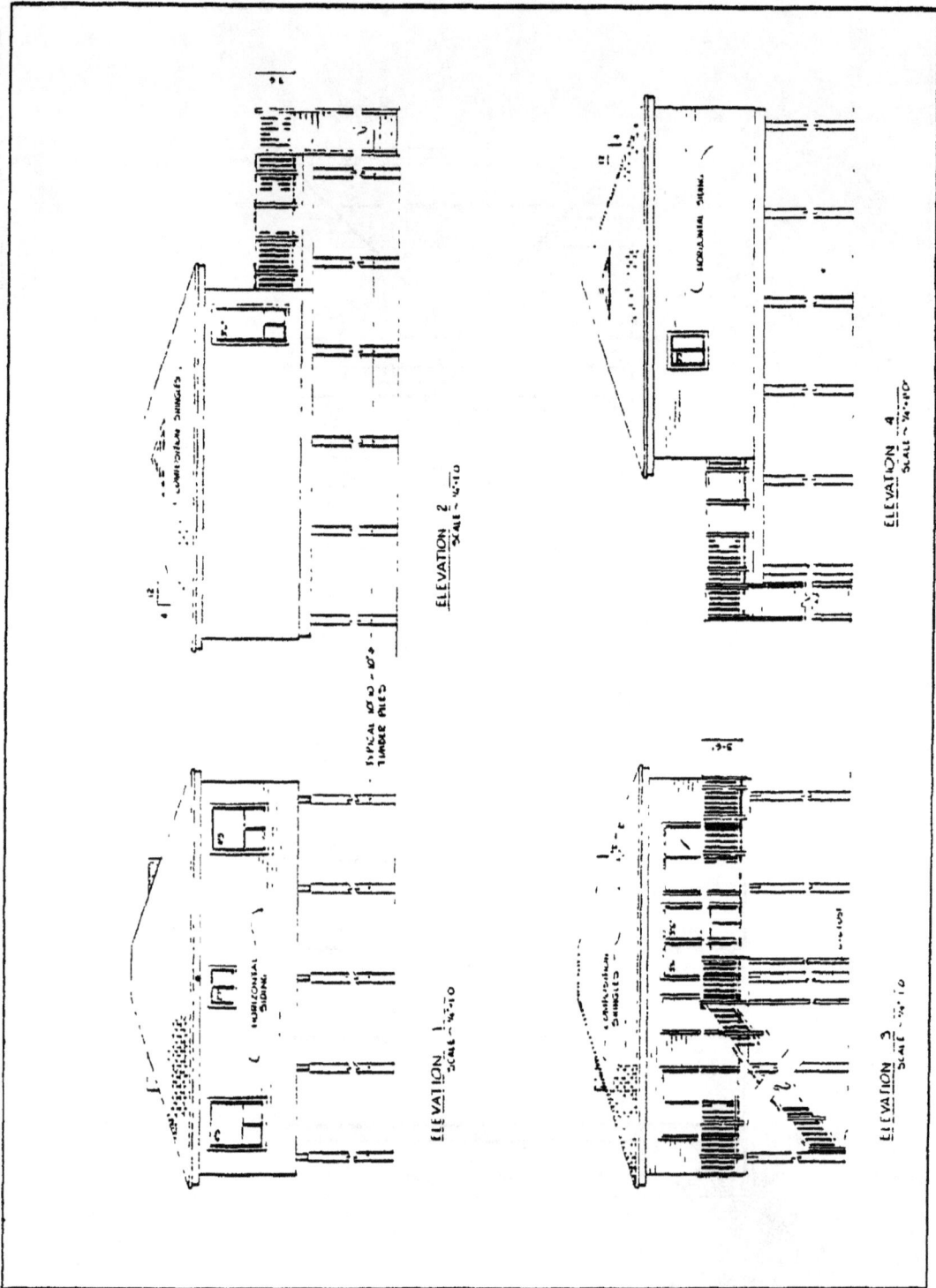

ELEVATION 1
SCALE ~ 1/4"=1'0

ELEVATION 2
SCALE ~ 1/4"=1'0

ELEVATION 3
SCALE ~ 1/4"=1'0

ELEVATION 4
SCALE ~ 1/4"=1'0

POOR QUALITY

The third step is to determine the beam size and connector requirement for attaching the beams to the piles. Section II. F. Beams of Chapter III indicates the minimum beam size to be 3 - 2x12 number 2 surface dry Southern Pine or Douglas Fir timbers. The connector loads are determined from Table 4-7, Figure 4-33. The load for this one story residence is 1600 pounds, providing that the beams bear fully on the piles. This load requires the beam to be connected to the pile with 2 - 1/2" diameter galvanized bolts spaced 3-1/2s oc.

FIGURE 4-33

*TABLE 4-7
MINIMUM BOLT SPACING 3-1/2", MINIMUM EDGE DISTANCE 2"
CONTINUOUS BEAM TO PILE OR POST CONNECTION LOADS AND BOLT REQUIREMENTS

	110 mph maximum	120 mph maximum
	8'-0" Maximum Span	
one story	2 - 1/2" diam. bolts 1600 pounds	2 - 1/2" diam. bolts 1930 pounds
two story	2 - 5/8" diam. bolts 2000 pounds	2 - 3/4" diam. bolts 3860 pounds
	10'-0" Maximum Span	
one story	2 - 5/8" diam. bolts 2000 pounds	2 - 3/4" diam. bolts 2410 pounds
two story	3 - 3/4" diam. bolts 2500 pounds	3 - 7/8" diam. bolts 4820 pounds

*Table 4-7 is based on beams fully bearing on piles or post. If beams are not fully bearing, the connector loads increase to 6000 pounds for one story buildings and 9000 pounds for two story buildings. In addition, when the connectors are installed, the bolt sizes and configuration must strictly adhere to the approved plans.

The fourth step is to determine the floor joist size and connector requirements. Section VI of this Chapter indicates the minimum floor joist size to be a 2" x 8" #3 grade Southern Pine or Douglas Fir timbers. The connector load can be determined from Table 4-9, Figure 4-34. Since the framing plan, Figure 4-32, indicates a 1'-6" overhang and 16" spacing, the 8'-0" span with 4'-0" cantilever column will be used, resulting in a connector load of 470 pounds.

FIGURE 4-34

TABLE 4-9
JOIST TO BEAM CONNECTOR LOADS

Spacing	8'-0" Span	8'-0" Span 4'-0" Cantilever
16" o.c.	270 Pounds	470 Pounds
24" o.c.	400 Pounds	600 Pounds
Spacing	10'-0" Span	10'-0" Span 4'-0" Cantilever
16" o.c.	335 Pounds	535 Pounds
24" o.c.	500 Pounds	700 Pounds

A number of connector manufacturers publish information relative to the load that various connectors will support. A typical example is shown in Figure 4-35. This connector example is an extract from The Panel Clip Company Brochure. Appendix A of this manual contains a number of connector manufacturers' publications. Entering the table for joist and purlin hangers, we find that a U26 connector is required and that it must be attached with 4 - 10d nails to the joist and beam.

FIGURE 4-35
PANEL CLIP JOIST HANGER TABLE

JOIST AND PURLIN HANGERS

U   ECONOMY JOIST HANGER (JOIST CLIP)

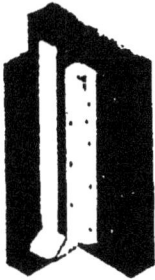

Product Application: Provides faster installation with the speed nail, cuts labor costs.

Joist Sizes: 2 x 4 to 2 x 14

Steel Designation: 18 ga. ASTM A-525 galvanized steel

Specify or Order: Panel Clip U—Economy Joist Hanger.

* Slightly different version & size supplied east of Continental Divide. Manufactured in Farmington, MI and Hayward, CA.

STK NO	JOIST SIZE	MATERIAL	DIMENSIONS (inches)				NAIL SCHEDULE		DESIGN LOADS (lbs)				
			A	D	H	W	HEADER	JOIST	NORMAL		MAXIMUM		UPLIFT
									10d	16d	10d	16d	
U24	2 x 4 to 2 x 6	18 ga. gal	1⅛	2	3	1⁹/₁₆	4-16d	2-10d	470	610	530	755	310
U26	2 x 8 to 2 x 10	18 ga. gal	1⅛	2	4⅛	1⁹/₁₆	6-16d	4-10d	840	810	810	1010	625
U210	2 x 10 to 2 x 14	18 ga. gal	1⅛	2	7³/₁₆	1⁹/₁₆	10-16d	6-10d	1175	1280	1330	1420	940
U26-2	(2) 2 x 6 to (2) 2 x 10	16 ga. gal			5¼	3³/₁₆	—		—	—	—	—	—
U210-2	(2) 2 x 10 to (2) 2 x 14	16 ga. gal			7¹¹/₁₆	3³/₁₆	—		—	—	—	—	—

4 U26 and U210 are manufactured from 20 ga. steel in Canada

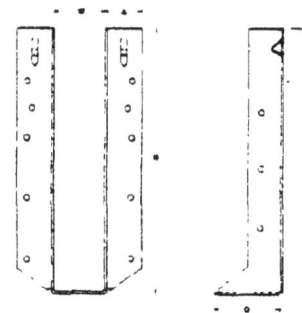

It should be noted that this manufacturer and others indicate that the connectors are to be galvanized, but does not indicate the amount of galvanized coating used per surface area. If this connector is to be used, the builder and inspector should ensure that the connector has a minimum of 1 ounce galvanizing per square foot of connector surface area.

The fifth step is to design the walls. The walls of this building are to be wood studs spaced 16" oc. Section IX. B. Wood Studs indicates the stud to be a minimum 2" x 4" stud or number 3 grade Southern Pine or Douglas Fir timber. Before we can determine the connector loading, we must look at the roof framing plan Figure 4-31, and the elevations Figure 4-32, to determine the roof span and the presence of an overhang. From the roof framing plan, we find the maximum roof span to be 35'-0" with a 1'-0" overhang on all sides, and a rafter spacing of 16" oc. Using these values, we enter Table 4-13, Figure 4-36 and determine the connector load to be 500 pounds.

FIGURE 4-36

TABLE 4-13
CONNECTOR LOADS (LBS) FOR
VARIOUS RAFTER SPANS
AND OVERHANGS

| WIDTH | 110 MPH VELOCITY | | | 120 MPH VELOCITY | | |
| | OVERHANG | | | OVERHANG | | |
	0'	2'	4'	0'	2'	4'
20'	215	285	360	260	345	435
24'	260	330	400	310	400	485
30'	320	395	465	390	475	565
32'	345	415	485	415	500	585
40'	430	500	575	520	605	690

*Loads based on 16" oc rafter spacing, 9 psf dead load and rafter slope between 0° - 40°. For 24" spacing, increase loads by 50%.

Once again manufacturer's publications can be consulted to find a connector to transfer the load from the wall to the floor structural system. It appears a strap connector is best suited for this design. In attaching the connectors from the studs in the east and west walls, the builder must install the straps so that the strap is not nailed into the end of the floor joist. Figure 4-37 is a typical manufacturer's table indicating the various size strap connectors and loads. The Simpson Strong Tie Company connector would be an FHA 9 and the strap would be attached to the stud and joist with 4 - 16d nails in each member.

FIGURE 4-37
SIMPSON STRONG TIE STRAP CONNECTOR

ST/FHA/MST/HST
STRAP TIES

APPLICATIONS: Where plates or soles are cut, ridge ties, wall intersections, truss plates. Specials made to order.

Single or double row bolt patterns for HST. Double row design values are for glulam beam applications.

Model No.	DIMENSIONS Material	Width	Length	FASTENERS Nails	Bolts	DESIGN LOADS Nails	BOLTS Single Shear	Double Shear
ST292	20 ga galv	2 1/16"	9 5/16"	12-16d	—	805	—	—
ST2122	20 ga galv	2 1/16"	12 13/16"	16-16d	—	1070	—	—
ST2115	20 ga galv	3/4"	16 5/16"	10-16d	—	450	—	—
ST2215	20 ga galv	2 1/16"	16 5/16"	20-16d	—	1210	—	—
ST6215	16 ga galv	2 1/16"	16 5/16"	20-16d	—	1340	—	—
ST6224	16 ga galv	2 1/16"	23 5/16"	28-16d	—	1875	—	—
ST6236	14 ga galv	2 1/16"	33 13/16"	40-16d	—	2520	—	—
ST9	16 ga galv	1 1/4"	9"	8-16d	—	535	—	—
ST12	16 ga galv	1 1/4"	11 5/8"	10-16d	—	670	—	—
ST18	16 ga galv	1 1/4"	17 3/4"	14-16d	—	935	—	—
ST22	16 ga galv	1 1/4"	21 5/8"	18-16d	—	1205	—	—
9	12 ga galv	1 1/2"	9"	8-16d	—	535	—	—
.12	12 ga galv	1 1/2"	12"	8-16d	—	535	—	—
FHA18	12 ga galv	1 1/2"	18"	8-16d	—	535	—	—
FHA24	12 ga galv	1 1/2"	24"	8-16d	—	535	—	—
FHA30	12 ga galv	1 1/2"	30"	8-16d	—	535	—	—
MST27	12 ga galv	2 1/16"	27"	15-16d ea end	2-1/2" ea end	2006	1612	3225
MST37	12 ga galv	2 1/16"	37 1/2"	21-16d ea end	3-1/2" ea end	2808	2418	4840
MST48	12 ga galv	2 1/16"	48"	25-16d ea end	4-1/2" ea end	3345	3224	6450
HST2	5/16"	2 1/2"	21 1/4"	—	6-5/8"	—	3765	7530
HST5	5/16"	5"	21 1/4"	—	12-5/8"	—	7537	15075
HST3	1/4"	3"	25 1/2"	—	6-3/4"	—	5435	10875
HST6	1/4"	6"	25 1/2"	—	12-3/4"	—	10875	21750

ACCEPTED—See Research Recommendation No. 1746 of the International Conference of Building Code Officials (Uniform Building Code). FHA, MST and HST series straps are UBC code approved nail and/or bolt values.

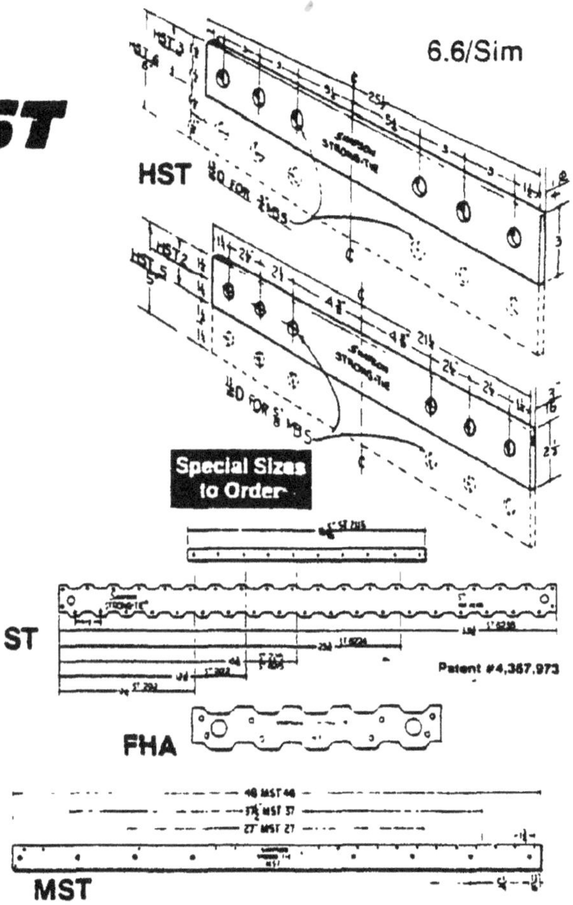

6.6/Sim
HST
Special Sizes to Order
ST
Patent #4,367,973
FHA
MST

The sixth step is to design the corner bracing. Table 4-14, Figure 4-38 indicates the nail requirement for both particleboard and plywood. For our purpose, we would select the 32'-0" building width for the north and south wall and 40'-0" for the east and west walls.

FIGURE 4-38

TABLE 4-14
NAIL SIZE AND SPACING
FOR CORNER SHEATHING

| | 110 mph maximum | | | | 120 mph maximum | | | |
| | 1 Particleboard | | 2 Plywood | | 1 Particleboard | | 2 Plywood | |
Width	Size	Spacing	Size	Spacing	Size	Spacing	Size	Spacing
20	10d	6" oc	10d	6" oc	10d	6" oc	10d	6" oc
24	10d	6" oc	10d	6" oc	10d	6" oc	10d	6" oc
30	10d	6" oc	10d	6" oc	10d	6" oc	10d	6" oc
32	10d	6" oc	10d	6" oc	10d	6" oc	10d	6" oc
40	10d	4" oc	10d	4" oc	10d	4" oc	10d	4" oc

NOTES:
1. Minimum Grade particleboard - 1/2" - 2M-W
2. Plywood Grade - 15/32" C-D Sheathing, C-C Exterior and Panel
   Siding

It is assumed for this design that particleboard will be used. The
use of particleboard requires 10d nails for the corner sheathing on
all four sides with the nails spaced 6" on center on the north and
south walls and 4" on the east and west walls.

The seventh step is the sizing of the roof structural system and the
connection of the roof system to the walls. The maximum span of the
rafters is 16'-0" with a 16" oc spacing. Using Table 4-15, Figure
4-39, we determine that a 2" x 8" number 2 grade Southern Pine
or Douglas Fir timber is required.

FIGURE 4-39

TABLE 4-15
RAFTER SIZE BASED ON SPAN @ 16" o.c.

Rafter Size	Maximum Span #3 Grade	Maximum Span #2 Grade
2 x 6	10'	12'
2 x 8	12'	16'
2 x 10	16'	20'
2 x 12	20'	24'

The framing plan indicates the roof span reduces from a maximum of
16 feet at the center to approximately 2 feet at the corners.
The rafter size could be reduced to 2" x 6" when the span does not
exceed 12'-0" This reduction may introduce framing problems and as
a result may not be practical.

The connector load can be obtained from Table 4-13, Figure 4-40. The width of the building in the east-west direction is 35 feet with a 2 foot overhang. Entering Table 4-13 at 40' with 2'-0" overhang, we find a connector load of 500 pounds. The 32 foot span with 2 foot overhang in the north-south direction results in a 500 pound load.

FIGURE 4-40

TABLE 4-13
CONNECTOR LOADS (LBS) FOR
VARIOUS RAFTER SPANS
AND OVERHANGS

WIDTH	110 MPH VELOCITY OVERHANG			120 MPH VELOCITY OVERHANG		
	0'	2'	4'	0'	2'	4'
20'	215	285	360	260	345	435
24'	260	330	400	310	400	485
30'	320	395	465	390	475	565
32'	345	415	485	415	500	585
40'	430	500	575	520	605	690

*Loads based on 16" oc rafter spacing, 9 psf dead load and rafter slope between 0° - 40°. For 24" spacing, increase loads by 50%.

Manufacturer's literature is consulted to determine the type connector and nails required to make the connection. Figure 4-41 is a typical example of the information available. The Ty-Down Senior Connector is selected from the TECO Products Catalog as it has a safe working load of 650 pounds. The special 1-1/2" long banded nail furnishing with the connector must be used to fasten the connector to the rafter and the stud.

FIGURE 4-41
TECO TY-DOWN ANCHOR SPECIFICATIONS

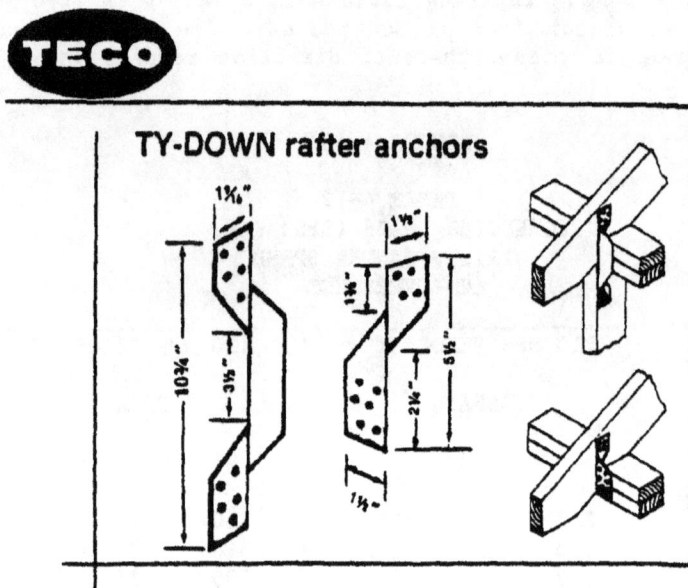

TY-DOWN rafter anchors

where used

For the most efficient framing of a structure, rafters or trusses should be anchored securely to wall studs rather than to plates alone. TECO Ty-Down Rafter Anchors accomplish this and provide increased resistance to uplift due to winds. Ty-Down, Srs. are used where rafters or trusses fall in line with the stud. Ty-Down, Jrs. are used at points where rafters or trusses are not in line with the studs. They can also be used separately as a tie-down device.

description

TECO Ty-Downs are precision manufactured from 18 gauge zinc coated steel. Ty-Down, Senior—10¾" long; flanges 1⅝" wide; two small flanges each have 5 nail holes; Recommended safe working value— 650# (Hemlock-470#). Ty-Down, Junior—5½" long; two flanges 1½" wide; one flange longer than other to reach second plate where desired; shorter flange has 4 nail holes, longer one has 6; Recommended safe working value—520# (Hemlock-375#). Special 0.133" diameter 1½" long nails (square barbed) are packed in each carton.

Meet HUD (FHA) Construction Standards.

Step seven is to select the ridge tie strap. The design load for this strap can be determined from Table 4-13, Figure 4-40. The sum of the maximum span of two abutting rafters is 32 feet with a 1'-0" overhang on each side and a 16 inch space. Entering Table 4-13, the load on the strap is determined to be 500 pounds.

Consulting a manufacturer's catalog, we can select the plate strap size. Figure 4-42 is an example of a plate strap table from the United Steel Products Company Catalog. From this table, we determine a S-65 strap with a 590 pound design load will be used with 12 - 10d nails (5 each) rafter.

FIGURE 4-42
KANT SAG STRAP CONNECTOR

STRAPS; TRUSS PLATES

PLATE STRAP
S SERIES

Design features: Helps resist movement caused by earthquakes, tornados and hurricanes.

Materials: 14- and 16-ga galvanized steel.

Loads: 120-to-470

ICBO #3573
BOCA #79-11
SBCCI #8003

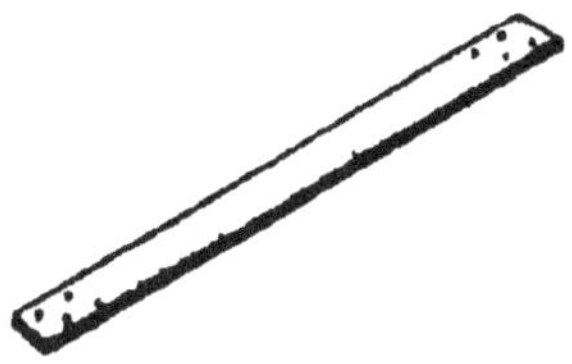

KANT-SAG.

STOCK NO.	DESCRIPTION	STEEL GA	UNITS PER CASE	CASE WT (LB)	NAIL SCHEDULE	DESIGN LOAD (LB)
S-60	1⅜x4 in	14	50	6	4-10D	120
S-61	1⅜x6 in	14	50	9	6-10D	235
S-62	1⅜x8 in.	14	50	12	8-10D	355
S-63	1⅜x9 in.	14	50	14	9-10D	355
S-64	1⅜x10 in	14	50	16	10-10D	470
S-65	1⅜x12 in.	14	50	18	12-10D	590
S-66	1⅜x14 in.	14	50	21	8-10D	470
S-67	1⅜x16 in.	14	50	24	8-10D	470
S-68	1⅜x18 in.	14	50	28	8-10D	470
S-69	1⅜x20 in.	14	50	31	8-10D	470
S-70	1⅜x24 in.	14	50	37	8-10D	470
S-71	1⅜x36 in	14	50	55	8-10D	470

STOCK NO.	DESCRIPTION	STEEL GA	UNITS PER CASE	CASE WT (LB)	NAIL SCHEDULE	DESIGN LOAD (LB)
S-80	1⅜x4 in	16	50	5	4-10D	120
S-81	1⅜x6 in.	16	50	7	6-10D	235
S-82	1⅜x8 in.	16	50	10	8-10D	355
S-83	1⅜x9 in.	16	50	11	9-10D	355
S-84	1⅜x10 in	16	50	13	10-10D	470
S-85	1⅜x12 in	16	50	14	12-10D	590
S-86	1⅜x14 in.	16	50	17	8-10D	470
S-87	1⅜x16 in.	16	50	19	8-10D	470
S-88	1⅜x18 in.	16	50	23	8-10D	470
S-89	1⅜x20 in.	16	50	25	[illegible]	470
S-90	1⅜x24 in.	16	50	30	8-10D	470
S-91	1⅜x36 in.	16	50	44	8-10D	470

Since we do not have a gable roof no further structural consideration must be given to this design. If the load bearing wall in detail B on Figure 4-30 was designed to reduce the rafter span, the connections of the braces, and purlin to the rafter and brace to the top plates and studs of the load bearing wall would require a connector of adequate size to resist the uplift wind forces. In addition, a connector would be required between the stud, bottom plate and floor structural system.

Once we have the building designed the plans should be submitted to the Building Inspection Department for approval and receipt of a building permit.

CHAPTER V

PERMIT APPLICATION AND INSPECTION

I.   GENERAL

When the drawings are complete and the builder or owner is ready to
start construction, a building permit application is filed with the
building inspection department. In a number of building
departments, the permit application is a part of the building permit
(Figure 5-1). The information required on the permit application
normally relates to identification of the owner, contractor and, if
applicable, architect or engineer, and information about the the
building to be constructed.

FIGURE 5-1
BUILDING PERMIT

BUILDING PERMIT

Jurisdiction of_____

Applicant to complete numbered spaces only.

JOB ADDRESS

| 1 LEGAL DESCR. | LOT NO. | BLK | TRACT | □ SEE ATTACHED SHEET |

| 2 OWNER | MAIL ADDRESS | ZIP | PHONE |

| 3 CONTRACTOR | MAIL ADDRESS | PHONE | REGISTRATION NO. |

| 4 ARCHITECT OR DESIGNER | MAIL ADDRESS | PHONE | REGISTRATION NO. |

| 5 ENGINEER | MAIL ADDRESS | PHONE | REGISTRATION NO. |

| 6 LENDER | MAIL ADDRESS | BRANCH |

7 USE OF BUILDING

8 Class of work: □ NEW □ ADDITION □ ALTERATION □ REPAIR □ MOVE □ REMOVE

9 Describe work:

---

10 Valuation of work: $

| PLAN CHECK FEE | | PERMIT FEE | |

SPECIAL CONDITIONS

Type of Const	Occupancy Group	Division
Size of Bldg. (Total) Sq. Ft.	No. of Stories	Max. Occ. Load
Fire Zone	Use Zone	Fire Sprinklers Required □ Yes □ No
No. of Dwelling Units	OFFSTREET PARKING SPACES: Covered	Uncovered

| APPLICATION ACCEPTED BY | PLANS CHECKED BY | APPROVED FOR ISSUANCE BY |

Special Approvals	Required	Received	Not Required
ZONING			
HEALTH DEPT.			
FIRE DEPT.			
SOIL REPORT			
OTHER (Specify)			

NOTICE

SEPARATE PERMITS ARE REQUIRED FOR ELECTRICAL, PLUMBING, HEATING, VENTILATING OR AIR CONDITIONING.
THIS PERMIT BECOMES NULL AND VOID IF WORK OR CONSTRUCTION AUTHORIZED IS NOT COMMENCED WITHIN 6 MONTHS, OR IF CONSTRUCTION OR WORK IS SUSPENDED OR ABANDONED FOR A PERIOD OF 1 YEAR AT ANY TIME AFTER WORK IS COMMENCED.

I HEREBY CERTIFY THAT I HAVE READ AND EXAMINED THIS APPLICATION AND KNOW THE SAME TO BE TRUE AND CORRECT. ALL PROVISIONS OF LAWS AND ORDINANCES GOVERNING THIS TYPE OF WORK WILL BE COMPLIED WITH WHETHER SPECIFIED HEREIN OR NOT. THE GRANTING OF A PERMIT DOES NOT PRESUME TO GIVE AUTHORITY TO VIOLATE OR CANCEL THE PROVISIONS OF ANY OTHER STATE OR LOCAL LAW REGULATING CONSTRUCTION OR THE PERFORMANCE OF CONSTRUCTION.

SIGNATURE OF CONTRACTOR OR AUTHORIZED AGENT   (DATE)

SIGNATURE OF OWNER (IF OWNER BUILDER)   (DATE)

WHEN PROPERLY VALIDATED (IN THIS SPACE) THIS IS YOUR PERMIT

PLAN CHECK VALIDATION   CK.   M.O.   CASH   PERMIT VALIDATION   CK.   M.O.   CASH

II. PLAN REVIEW

Most building departments require plans to be submitted along with the building permit application. The purpose of submitting the plans is to allow the building department to determine if the design complies with the adopted codes. A review of the plans in an office environment allows a more complete review and minimizes the number of construction changes required in the field for code compliance.

In coastal construction, plan review is even more important because of the number of additional requirements for hurricane resistance construction. To assist the building departments in this review, a plan review check list for one and two family dwellings and small commercial buildings is provided (Figure 5-2). This check list identifies the critical structural elements in coastal construction that should be reviewed by the plan reviewer. In addition, the plan reviewer check list is designed to communicate to the inspector information such as the sheet of the plans details and page of the specifications critical information is located that must be verified in the field. The plan examiner has space for short remarks that may further assist the inspector in the field. It should be emphasized that the plan examiner should do more than simply assure that the plans are in accordance with the basic standards. He should also communicate clearly to the field inspector the information on the plans and in the specifications that must be verified for correctness in the field by the inspector.

FIGURE 5-2
PLAN REVIEW CHECK LIST FOR ONE AND TWO FAMILY DWELLINGS
AND SMALL COMMERCIAL BUILDINGS

	PLAN REVIEW		FIELD INSPEC		
		Plan Review Remarks	OK	C	N/A
I.	General				
1.1	Design Wind Velocity				
1.2	Building Location				
1.3	Soil Information				
1.4	Flood Plain Elevation				
2.	Foundation				
2.1.1	Footing Excavation				
2.1.2	Footing Design				
2.1	Foundation Wall				
2.2.2	Foundation Wall to Footing Connection				
2.2.3	Elevation of Lowest Opening				
2.3.1	Pile Layout				
2.3.2	Pile Grade and Species				
2.3.3	Pile Treatment				
2.3.4	Pile Length				
2.3.5	Penetration Blows/Ft				
2.3.6	Hammer Size				
2.3.7	Cut-Off Elevation				
2.3.8	Breakaway Walls				
3.	First Floor Beams				
3.1	Beam Grade and Species				
3.2	Beam Treatment				

	Plan Review Remarks	OK	C	N/A
3.3 Beam Size				
3.4 Beam Span				
3.5 Beam Bearing				
3.6 Beam to Pile Connection				
3.7 Corrosion Protection				
4. Floor Joists				
4.1 Grade and Species				
4.2 Treatment				
4.3 Size				
4.4 C.C. Spacing				
4.5 Maximum Span				
4.6 Bearing				
4.7 Beam to Joist Connectors				
4.8 Corrosion Protection				
4.9 Cutting and Notching				
5. Floor Sheathing				
5.1 Grade				
5.2 Thickness				
5.3 Nailing				
6. Walls				
6.1 Stud Grade and Species				
6.2 Stud Size				
6.3 C.C. Spacing				
6.4 Stud to Sole Plate Connector				
6.5 Stud to Top Plate Connector				

	Plan Review Remarks	OK	C	N/A
6.6 Wall to Floor Connector				
6.7 Header Size				
6.8 Cutting and Notching				
7. Wall Sheathing				
7.1 Grade				
7.2 Thickness				
7.3 Nailing				
8. Roof Framing				
8.1 Grade and Species				
8.2 Size				
3 C.C. Spacing				
8.4 Maximum Span				
8.5 Wall to Rafter Connectors				
8.6 Ceiling Joist to Rafter				
8.7 Ridge Tie Strap				
8.8 Gable End Bracing				
9. Roof Sheathing				
9.1 Grade				
9.2 Thickness				
9.3 Nailing				
10. Roof Coverings				

Once the plan examiner has properly executed the suggested check sheet, the same check sheet becomes a useful tool for the field inspector.

The plans submitted for the Alicia Residence (Figures 5-3 to 5-7) will be reviewed in accordance with the check list. When reviewing the plans for compliance with coastal construction requirements, the reviewer need only to compare the size of the structural elements on the plans with the various design tables contained in Chapter IV. Since this almost repeats the design section in Chapter IV, this comparison is not made in Chapter V. Figure 5-8 is a completed plan check list of the Alicia Residence.

FIGURE 5-3
SITE PLAN ALICIA RESIDENCE

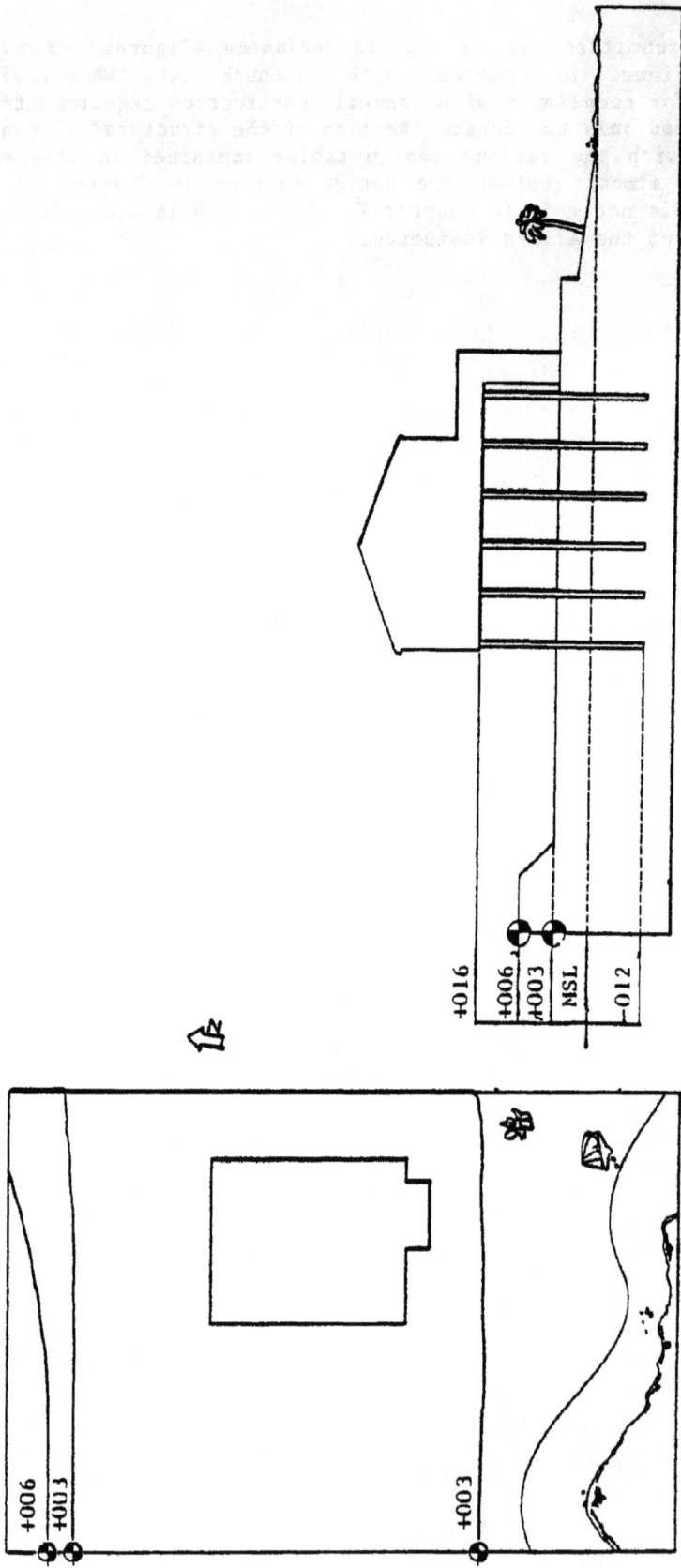

+016
+006
+003
MSL
-012

+006
+003

+003

FIGURE 5-4
ALICIA RESIDENCE

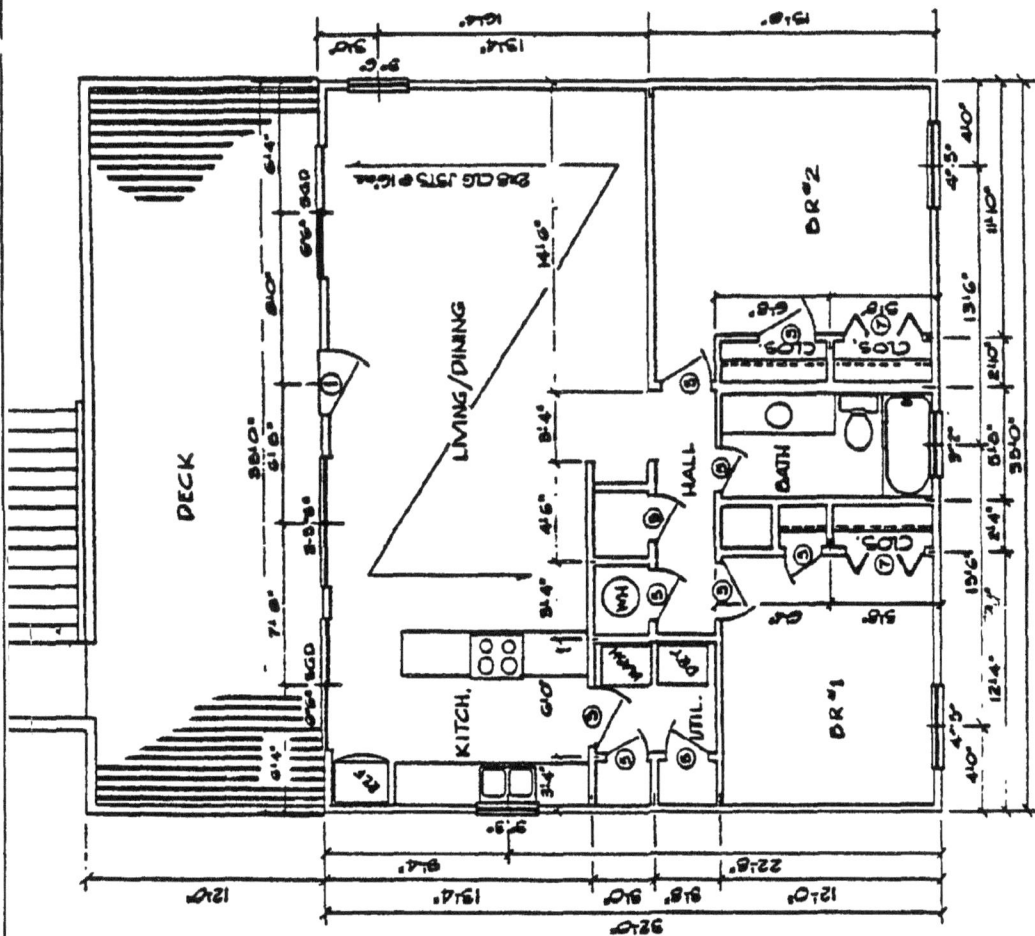

FIRST FLOOR PLAN
SCALE ~ 1/4" = 1'-0"

ALICIA RESIDENCE

FIGURE 5-5
ALICIA RESIDENCE

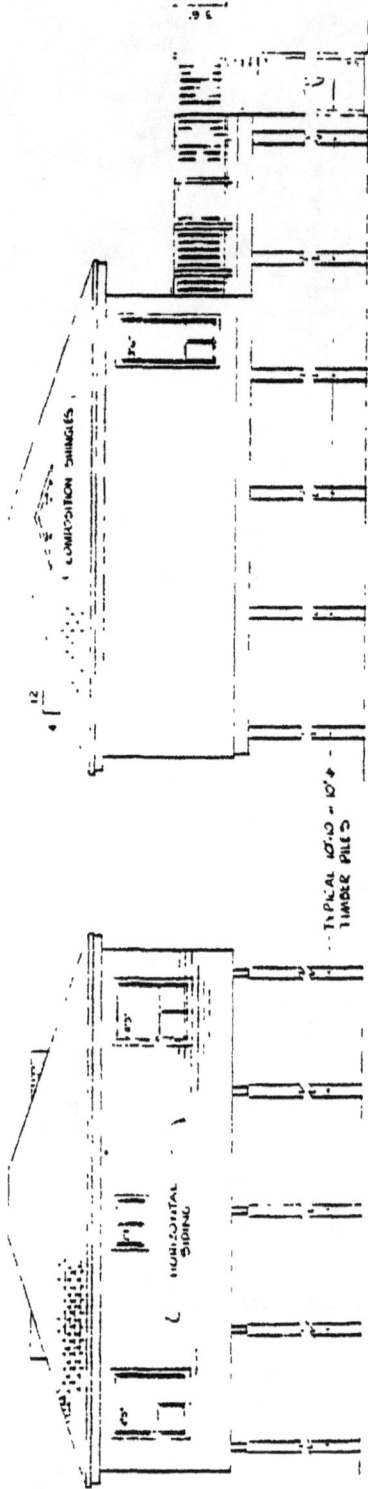

ELEVATION 2
SCALE ~ 1/4"=1'-0"

ELEVATION 4
SCALE ~ 1/4"=1'-0"

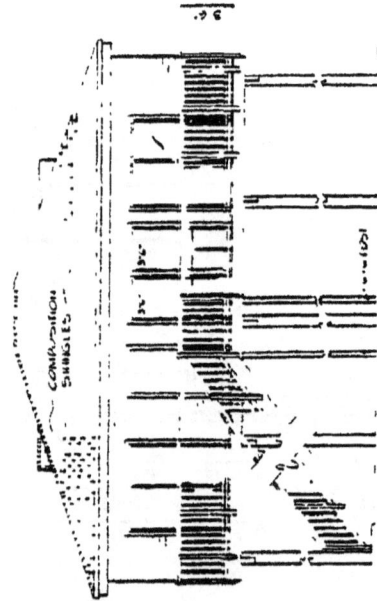

ELEVATION 1
SCALE ~ 1/4"=1'-0"

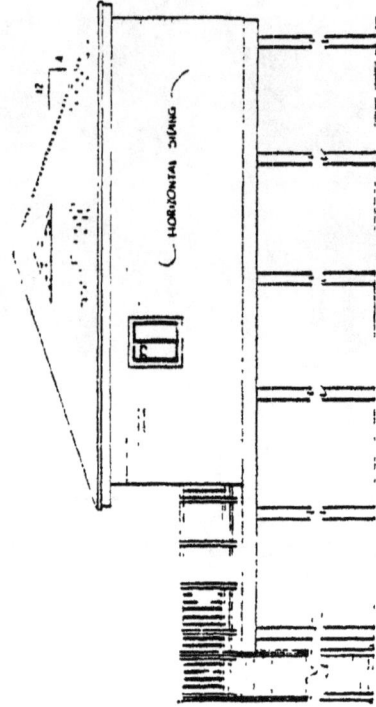

ELEVATION 3
SCALE ~ 1/4"=1'-0"

ALICIA RESIDENCE

POOR QUALITY

FIGURE 5-6
ALICIA RESIDENCE

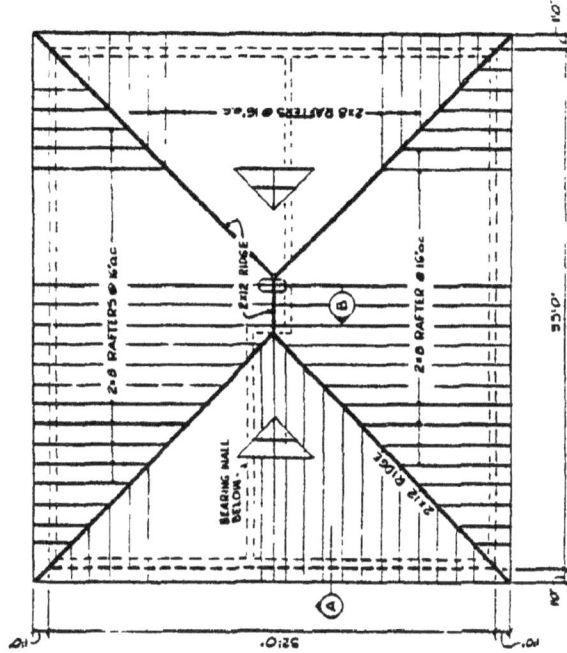

ROOF FRAMING PLAN
SCALE: ¼"·1'0"

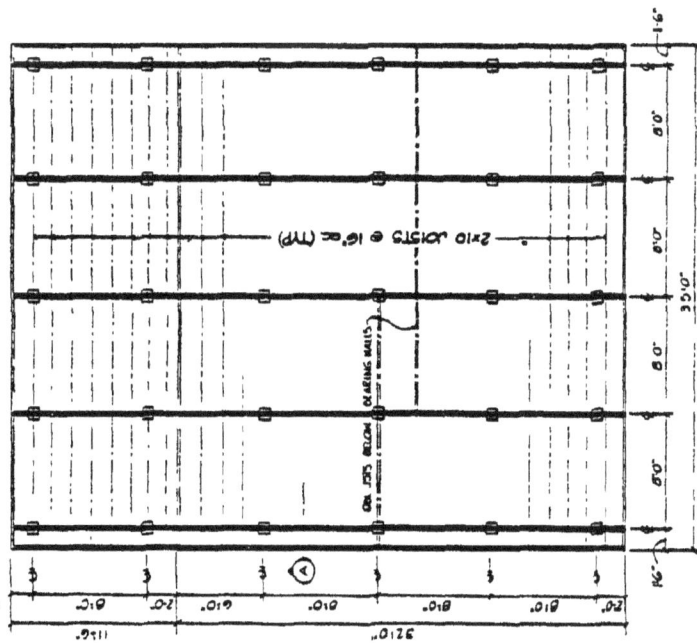

FIRST FLOOR FRAMING PLAN
SCALE- ¼"·1'0"

FIGURE 5-7
ALICIA RESIDENCE

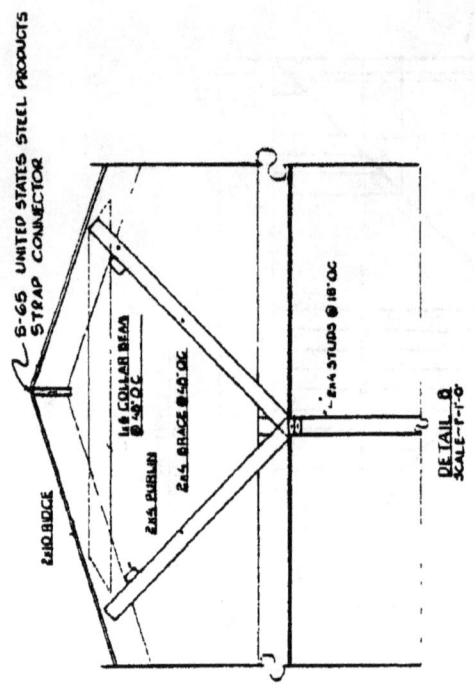

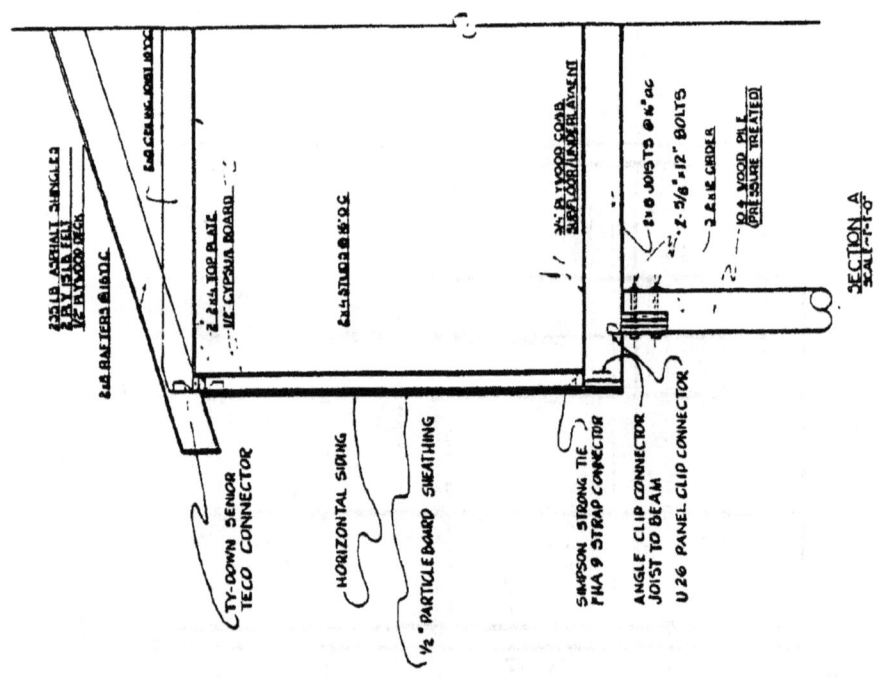

DETAIL B
SCALE-1"-1'-0"

SECTION A
SCALE-1"-1'-0"

FIGURE 5-8
COMPLETED PLAN CHECK LIST OF ALICIA RESIDENCE

PLAN REVIEW		FIELD INSPEC		
	Plan Review Remarks	OK	C	N/A
1. General				
1.1 Design Wind Velocity	110 mph			
1.2 Building Location	Reference site plan			
1.3 Soil Information	Finished grade - 3 ft above mean sea level - sandy course gravel			
1.4 Flood Plain Level	12'-0" Above mean sea level			
2. Foundation				
2.1.1 Footing Excavation				
2.1.2 Footing Design				
2.2.1 Foundation Wall				
.2.2 Foundation Wall to Footing Connection				
2.2.3 Elevation of Lowest Opening				
2.3.1 Pile Layout	Reference site plan and first floor framing plan			
2.3.2 Pile Grade and Species	Douglas Fir			
2.3.3 Pile Treatment	AWPB-C-3			
2.3.4 Pile Length	30 ft length, 10" tip diam.			
2.3.5 Penetration Blows/Ft	20 blows/ft. - last 1 ft.			
2.3.6 Hammer Size	700 Pounds			
2.3.7 Cut-Off Elevation	13 ft. above grade			
2.3.8 Breakaway Walls				
3. First Floor Beams				
3.1 Beam Grade and Species	#2 Surface Dry SP			
3.2 Beam Treatment				

13-5

		Plan Review Remarks	OK	C	N/A
3.3	Beam Size	3 - 2" x 12"			
3.4	Beam Span	8 ft.			
3.5	Beam Bearing	Full Bearing on All Piles			
3.6	Beam to Pile Connection	2 - 5/8" bolts - 3-1/2 oc			
3.7	Corrosion Protection	1 oz. Galv.			
4.	**Floor Joists**				
4.1	Grade and Species	#3 Southern Pine			
4.2	Treatment				
4.3	Size	2"x8"			
4.4	C.C. Spacing	16" o.c.			
4.5	Maximum Span	8" with 1'-6" Cantilever			
4.6	Bearing	1-1/2" Bearing - Solid Blocking at each support			
4.7	Beam to Joist Connectors	U26 Panel Clip, 4 - 10d Nails to each joist			
4.8	Corrosion Protection	All nails galvanized			
4.9	Cutting and Notching	1/4 (D) Ends - 1/6 (D) Outer Third of Span			
5.	**Floor Sheathing**				
5.1	Grade	32/16			
5.2	Thickness	5/8"			
5.3	Nailing	6" o.c. Face, 10" o.c. Intermediate 6d Spiral Thread			
6.	**Walls**				
6.1	Stud Grade and Species	#3 Southern Pine			
6.2	Stud Size	2"x4"			
6.3	C.C. Spacing	16" o.c.			
6.4	Stud to Sole Plate Connector	2 - 16d galv. nails			
6.5	Stud to Top Plate Connector	2 - 16d galv. nails			

PLAN REVIEW		FIELD INSPEC		
	Plan Review Remarks	OK	C	N/A
6.6 Wall to Floor Connector	FHA9 - Simpson Strong Tie Strap - 1/8" thick, 9" length - 8 - 16d galv. nails			
6.7 Header Size	2 - 2"x8" - #2SP 16d nails - 16" o.c.			
6.8 Cutting and Notching	25% Bearing (Notches) 40% Non-bearing (Notches)			
7. Wall Sheathing				
7.1 Grade	Particleboard - 2-M-W			
7.2 Thickness	1/2 Inch			
7.3 Nailing				
8. Roof Framing				
8.1 Grade and Species	#3 Southern Pine			
8.2 Size	2"x8"			
8.3 C.C. Spacing	16" o.c.			
8.4 Maximum Span	16 ft.			
8.5 Wall to Rafter Connectors	Ty-Down Senior TECO Connector 1-1/2" long special square barbed nail by TECO			
8.6 Ceiling Joist to Rafter	3 - 16d face nail			
8.7 Ridge Tie Strap	S-64 1-3/8"x8" Kant Strap 10 - 10d nails			
8.8 Gable End Bracing				
9. Roof Sheathing				
9.1 Grade	16/0			
9.2 Thickness	3/8"			
9.3 Nailing	6d angular nails, 6" oc, 12" oc			
10. Roof Coverings				

III. INSPECTION

The prime function and responsibility of the inspector is to assure that the work is in all respects in accordance with the plans and specifications approved by the plan reviewer. Many inspection departments require the builder to have a copy of the plans at the building site to ensure that the inspector has access to the plans during his inspections.

In some cases, the specifications for the project, or standard specifications included therein by reference, may establish definite tolerances over or under the exact measurements that will be accepted, and the inspector then has to verify that the work is within specified limits. On many phases of work, however, specific tolerances cannot be fixed, and intelligent judgment is required in interpreting such requirements as plumb, true, level and perfect. We will assume in this section that the inspector has the practical knowledge of grades of workmanship and general inspection requirements and concentrate on special inspection concerns in hurricane areas.

The inspector must assure himself that the principal centerlines, column lines, and the controlling overall dimensions and elevations are correct. Minor errors may exist but they should be within normal construction tolerances and not permitted to accumulate. It is important to remember that the inspector basically can only determine what is _visually_ acceptable. The competence and knowledge of the builder and suppliers is still mandatory to assure proper construction.

It is important that the inspector make clear to the builder at the outset the work that will be inspected and make sure that the initial portions of the work fulfill his expectations. It will invariably be found that the standards of accuracy established and enforced during the first few days of work will set the pattern for the rest of the work. The inspector must be consistent in the standards enforced. He must be reasonable, but he cannot be lenient in this respect.

A review of the plan review check list (Figure 5-8) and approved plans by the inspector prior to leaving the office allows the inspector to be familiar with the specific details he must pay close attention to during his inspection. It also allows the inspector to identify field changes that have not been communicated to the building department. In many cases field changes will result in delays as the inspector will not have the proper approved information when he arrives for his inspection.

Since structural integrity of a building is of primary importance in coastal construction, the footing and foundation, and framing inspections are the most important inspections to be made by the inspector. Detail recommendations for these inspections follow.

A. Footing and Foundation Inspection

1. Site Location

The first inspection required is conducted prior to any concrete being poured or piles being driven. The inspector should first verify that either the spread footing excavation or the location of the stakes for the foundation are located as indicated on the SITE PLAN.

16-5

2. Spread Footings

A conventional spread footing would normally only be approved by the building department if it was located behind sea walls or other areas not subject to flooding. Since this is normal conventional construction few special concerns for the inspector beyond conventional construction are encountered by the inspector. In our typical example, the building is located in an area where a spread footing would be prohibited.

3. Pile Inspection

(a) Prior to Pile Driving - In our typical example a pile foundation is to be constructed. The plan examiner has indicated on the check sheet the specific information that should be verified by the inspector.

During the first inspection, the inspector should assure the following before pile driving begins:

    (1)   Overall pile length
    (2)   Pile size
    (3)   Pile treatment (quality assurance marking)
    (4)   Grade of pile
    (5)   Soil conditions

As indicated on the check sheet, the plan examiner has determined that the penetration depth for our example must be 15 feet below grade and the pile cutoff elevation must be 13 feet above grade. From this information the inspector knows that the overall pile length must be a minimum of 30 feet, which allows a maximum of 2 feet for leveling.

The inspector should verify the actual pile size. In this example a 10 inch diameter column is indicated. Pile treatment is assured by an AWPB quality control marking. Since in this example the pile will not be submerged completely below the permanent water table, the applicable AWPA/AWPB standard acceptable is C-3. Such a quality marking is indicated in Figure 5-9.

FIGURE 5-9
PRESERVATIVELY TREATED LUMBER STAMP GRADE

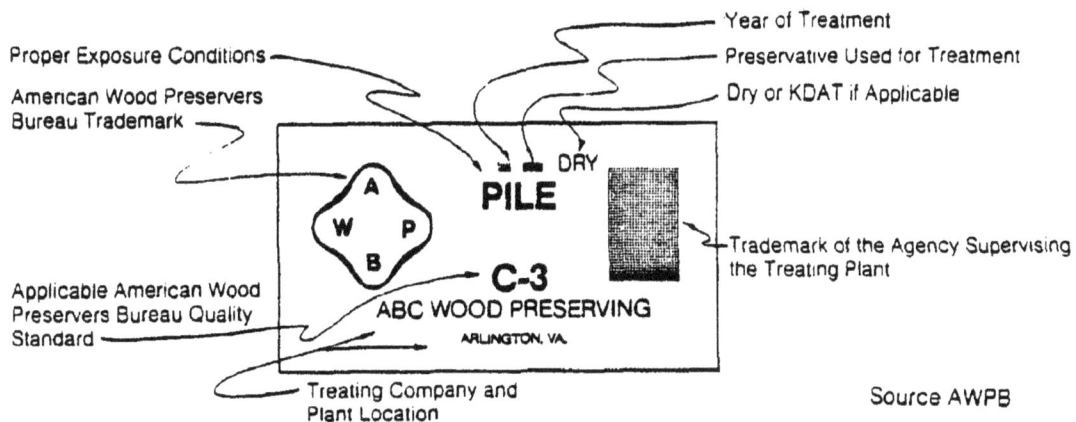

Preservatively Treated Lumber Stamp Grade

The plans review check sheet indicates the piles will be Douglas Fir. The inspector must verify that an appropriate Douglas Fir grade stamping is on the piles.

The inspector should finally visually verify the soil type is in accordance with the plans and specifications. In our example, the check sheet indicates sandy gravel.

(b) <u>During Pile Driving</u> - The inspection of pile driving is an extremely important phase of the inspector's duties. There is probably no other type of construction work that requires quick sound decisions to be made on the spot as frequently. Every inspector assigned to supervision of pile driving operations should familiarize himself thoroughly with all details of materials, equipment, and techniques used in pile driving and of application and limitations of the methods of evaluating safe bearing capacity.

The inspector should be present during the driving of each pile to be used for a permanent structure. Many jurisdictions do not have sufficient inspectors to allow time for inspectors full time during pile driving. This, however, is not sufficient reason not to recommend the full time inspection during this important phase of construction.

For our typical example, it can be estimated that each pile will take approximately one hour to drive, under normal conditions. It can be computed that approximately 30 hours should be allotted for inspection time on this one project because 30 piles are indicated.

The inspector should observe that the pile is handled without undue strain or shock, that the pile is set plumb in the leads, and that the pile-driver leads themselves are plumb.

The builder should assure the hammer is the appropriate weight for the pile being driven. The hammer should weigh as much as the pile being driven and preferably up to twice the weight of the pile.

Driving of piles through many types of soil can be facilitated by the use of jetting. Since the piles in a hurricane area may be subjected to uplift, this procedure should not be permitted unless under the supervision of a registered engineer.

The inspector must make sure that piles are driven to the minimum point of elevation specified. Since the piles are subjected to uplift simply driving to refusal cannot be accepted unless approved by the design engineer.

If refusal does occur, continued driving does little good and may in fact cause serious damage to the pile such as brooming the end or splitting the pile. If the lack of penetration seems to be due to an obstruction, it may be small enough so that 10 to 15 blows at reduced impact will dislodge or break through the obstruction to permit normal driving to continue.

The blows per foot of the last four feet of penetration should be documented by the inspector.

Breaking of a pile can usually be detected when the penetration suddenly increases for each blow of the hammer. If this occurs, the inspector should either require the pile to be pulled or a new pile driven.

(c) Final Pile Preparation - At the completion of the pile driving, the inspector should assure the subsequent driving of the piles has not caused adjacent previously driven piles to heave and the piles are not visually damaged. The cutoff elevation should be checked for appropriate elevation above grade.

B. Framing Inspection

1. Beam Framing

The inspector must verify that the size, span, grade species and bearing of the beams are in accordance with the approved plans and specifications.

For our example the plan examiner has indicated on the executed check sheet that the beams are 3 - 2x12 number 2 surface dry Southern Pine timbers, and the maximum clear span is approximately eight feet. The inspector should further verify that no visual damage has occurred to the beams during shipment and installation.

2. Beam Connection

The plan review check sheet indicates that the beam is connected to the pile with 2 - 5/8" diameter bolts spaced not more than 3-1/2" o.c. The bolts are further specified as being galvanized. The inspector must verify the bolt size, spacing, and corrosion protection are in accordance with the approved plans and specifications.

3. Floor Joists

The next item to verify during framing inspection is the floor joist framing. Our inspector's check list indicates the following for our example:

    Grade species - #3 grade Southern Pine
    Treatment - Not applicable
    Size - 2" x 8"
    Center to center spacing - 16" o.c.
    Maximum span - 8' with 1'-6" cantilever
    Bearing and blocking - 1-1/2" minimum - block at each support
    Beam to joist connectors - U26 panel clip with 4 - 10d nails
                                  to each joist
    Corrosion protection - Connector and nails galvanized
    Cutting and notching - Within conventional limitations

The inspector should assure he has adequate details such as the specifications on the specific connector given in the check sheet so he may verify compliance with documented information. The builder may modify many of the individual construction elements such as changing to an equal or better grade and specifies of wood or changing to another type of connector that is equivalent to what is required would certainly be permitted, but as earlier recommended such field changes should be communicated to the building department to assure proper information is given to the field inspector prior to the actual framing inspection.

4. Floor Sheathing

The inspector must verify that the floor sheathing is in accordance with our check list. The check list for this example indicates the following:

    Grade - 32/16 (panel identification index number) exterior
              grade
    Thickness - 5/8"
    Nailing - 6d number or spiral thread 6" o.c. face, 10" o.c.
              intermediate

5. Wall Framing

The inspector must next verify compliance of the wall framing with the check list. For our example problem, this includes the following:

    Grade species - #3 Southern Pine
    Stud size - 2" x 4"
    Center to center spacing - 16" o.c.
    Stud to sole plate connector - 2 - 16d galvanized nails
    Stud to top plate connector - 2 - 16d galvanized nails
    Wall to floor connector - FHA9-Strap - 1/8" thick, 9" length
                              8 - 16d nails
    Header size and nailing - 2 - 2" x 8" No. 2 Southern Pine
                              16d nails, 16" o.c. each edge
                              galvanized face nail to stud
                              4 - 16d nails each end to stud
    Cutting and notching - Within conventional limitations

6. Wall Sheathing

The areas requiring field inspection for the wall sheathing are listed as follows with the specific information for our example problem.

    Grade - Particleboard - 2-M-W
    Size - 1/2 inch
    Nailing - 10d nails - deformed shank schedule 6" o.c. north
              wall; 4" o.c. south wall galvanized nails

The inspector should require approximately a 1/16 inch gap between panel joints and assure a minimum of 3/8" of particleboard exists between the nail and the edge of the plywood.

7. Roof Framing

The critical items of roof framing that require field verification by the inspector with the specific information from the check sheet is listed as follows:

Grade species of rafters - #3 Southern Pine
Size - 2" x 8"
Center to center spacing - 16" o.c.
Maximum span - 16 feet
Wall to rafter connector - Ty Down Senior TECO Connector;
          1-1/2" long; special barbed
          nails furnished by TECO
Ceiling joist to rafter - 3 - 16d face nail
Ridge strap - S-65 - 1-3/8" x 8" Kant Sag Ridge Strap, 12 -
    10d nails
Gable end bracing - Not applicable (no gables)

8. Roof Sheathing

The field verifications required from the check list of our example problem are as follows:

Grade - 16/0 exterior plywood
Thickness - 3/8"
Nailing - 6d angular nails, 6" o.c. edges and 12" o.c.
   intermediate

9. Roof Coverings

The inspector must verify the shingles are supplied in accordance with specification. The nails must be corrosion resistant angular type nails.

Roof Coverings - 235 lb. asphalt shingles 12 gauge - 3/8" HD
     roofing nails - 4/36" shingle section

CHAPTER VI

COMMON DESIGN DEFICIENCIES

I.   INTRODUCTION

As previously noted, perhaps the greatest cause of building failures
in hurricanes is the lack of attention to details, especially
connections. While many designers, contractors and building officials
are familiar with the distribution of vertical loads in a given
structural system, the distribution of lateral loads are not as
evident. This chapter discusses the various structural systems used
to resist these lateral loads. The forces are traced from their point
of origin, through the building, and ultimately into the foundation.
Also discussed are various problems that have been associated with
these systems. The primary intent is to familiarize the professional
engineer with these problem areas so they may receive proper attention
during design and during the preparation of contract documents. This
section will also familiarize the contractor and building official
with the nature of lateral force resisting systems in order that they
may receive proper attention during construction and inspection.

II. PRIMARY LATERAL FORCE RESISTING SYSTEMS

Almost all ordinary rectangular buildings have one or more of the
lateral force resisting systems described below. The three primary
lateral force resisting systems that are commonly used are: 1) rigid
frames, 2) braced frames and 3) shearwalls. These three systems are
illustrated in Figure 6-1.

FIGURE 6-1
RIGID FRAME

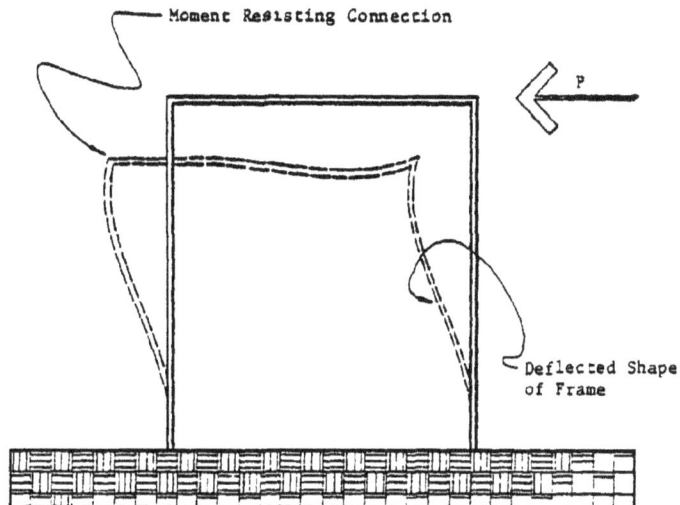

Moment Resisting Connection

P

Deflected Shape
of Frame

FIGURE 6-1 (Continued)
BRACED FRAME

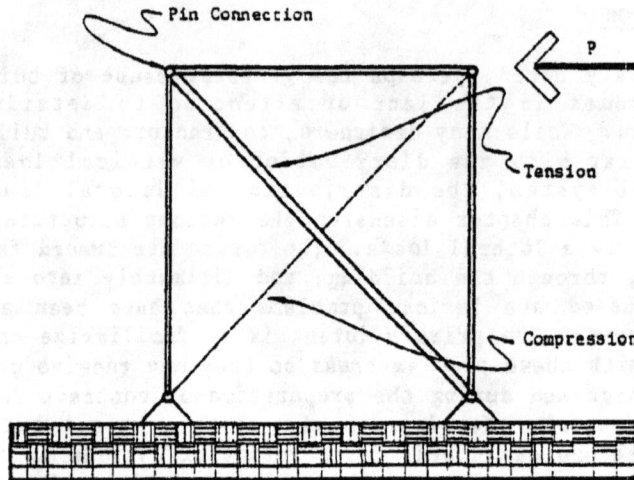

Pin Connection

P

Tension

Compression

FIGURE 6-1 (Continued)
SHEARWALL

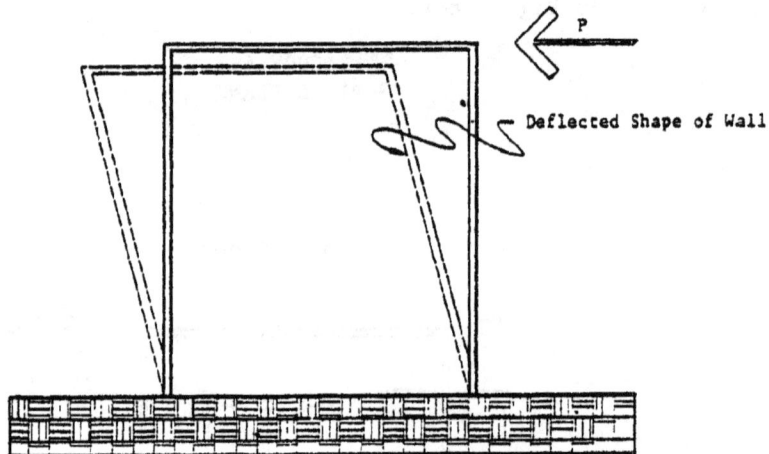

P

Deflected Shape of Wall

A. Rigid Frames

Rigid frames resist lateral loads by the bending of the columns and girders which make up the frame. Rigid frames are commonly used to resist lateral wind forces on metal frame buildings and concrete frame buildings. They are seldom used in buildings over 20 stories in height, unless used in combination with braced frames or shear walls. Other systems are more economical for higher buildings. A typical example of a rigid frame is the main frame on most pre-engineered metal buildings.

Since most rigid frame buildings are frequently fully engineered buildings, few failures are reported in these buildings that can be traced to the strength of the frames themselves. Some problems have been reported due to a lack of stiffness in the frames, which have resulted in cracking and serviceability problems of non-structural elements. These will be discussed in later sections. However, there are exceptions to this general statement concerning the strength of those buildings which would be classified as marginally engineered buildings.

Marginally engineered buildings are commonly designed by those who are not experienced structural engineers. Frequently design aids such load tables are used in sizing the structural members. Extreme caution should be exercised by those using such design aids unless they are familiar with the development and intended use of these items.

An example of this problem is often found in the design of strip shopping centers. Steel joist and joist girders supported on steel columns is a commonly used framing system (See Figure 6-2).

FIGURE 6-2
TYPICAL STEEL JOIST AND JOIST GIRDER FRAMING SYSTEM

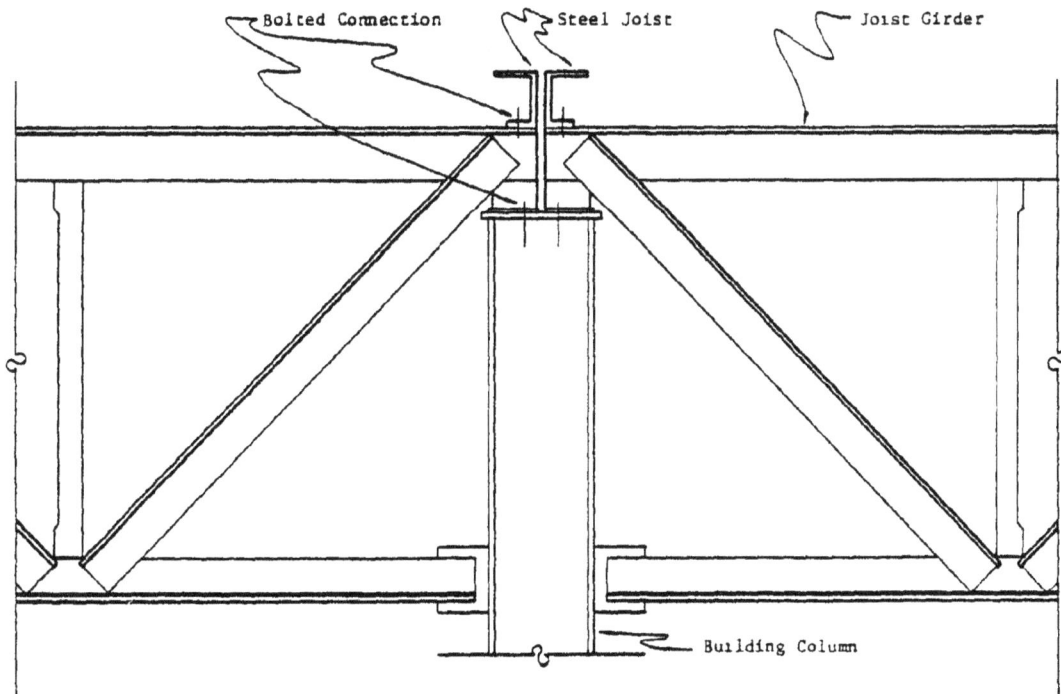

Designers tend to size these joists, joist girders, and columns for vertical loads only. Several problems result when this system is designed in such a manner:

1. No attention is given to wind uplift on the joists or joist girders. Since the members are designed based on load tables for gravity load only, unless otherwise specified by the designer, they are not designed for uplift loads produced by the wind. This can result in the buckling of their lower chord when subjected to high winds.

2. The bottom chord of the joist or joist girders is usually extended to the column and connected to it. This must be done to develop rigid frame action in order to stabilize the building against lateral loads, unless other lateral force resisting systems are present. However, a problem results from the fact that these members are usually specified on the basis of vertical gravity loads only, assuming simple support conditions. This arbitrary continuity, causes bending moments to develop in these members due to rigid frame action. These bending moments were not taken into account in the design of the member and some of these moments are present even under the dead load plus live load conditions. This can lead to the buckling of their lower chord and failure of their connections at the column. When this occurs, a main frame failure results.

3. Another problem is that the columns are commonly sized from load tables developed for axial loads only, assuming that their effective length is equal to their actual length. Thus, the bending capacity of the columns are not checked to verify their adequacy as an element of the rigid frame. Furthermore, the axial load capacity will be less than that considered unless the rotational restraints of the column ends are evaluated in order to determine the proper effective length to be used in determining the column size.

Similar problems can result when computer programs are used in the design of a building. A defective program or a user unfamiliar with the development, intended use, and limitations of the program can lead to defective design. This problem concerning the use of computers is not limited to rigid frames or hurricane design alone.

B. Braced Frames

Braced frames resist lateral forces by developing axial forces in the members as a vertical truss system. The brace system may use K-braces, X-braces, knee braces or other bracing patterns. This system can be economical up to about 60 stories in height and is typically used in mid-rise buildings in the form of a braced core. A brace system is usually used to resist forces which act perpendicular to the rigid frames in pre-engineered metal buildings. Such bracing usually takes the form of rod or cable bracing in the plane of the roof and sidewalls. The 1 x 4 wood let-in brace used in the corners of wood frame stud walls is yet another use of this system (however, the use of the corner brace is not recommended for use in non-engineered buildings which will be subjected to hurricane winds). Much like rigid frame buildings, buildings using the braced frame concept are usually fully engineered and failure of these frames are rare, even when the buildings are subjected to hurricane force winds. Stiffness considerations concerning the impact of frame deflection on nonstructural elements is less a problem with braced frames than with rigid frames since the brace frame system tends to have a greater stiffness due to its very nature.

C. Shearwalls

The shearwall system is a system where a wall is designed to function as a shear-resisting element carrying lateral loads. The deflection of the wall is predominately due to shear deformation rather than bending deformation, hence the name shearwall. However, the term is somewhat of a misnomer as far as its use in high rise buildings is concerned, since such shearwalls are usually quite slender. In this case, deflection is predominantly due to bending with only insignificant shear distortions.

This lateral force resisting system, in combination with rigid frames, is very common in high rise buildings up to about 70 stories. Another typical application includes buildings up to several stories where the bearing walls of masonry are used in conjunction with cast-in-place or precast concrete, steel, or wood floors and roofs. Yet another example is wood construction where sheathed wood stud walls are used.

High rise buildings using shearwall system rarely present problems when subjected to hurricane force winds. However, buildings having masonry bearing walls or wood frames have presented considerable problems. This is primarily because these buildings are either marginally engineered or non-engineered. Even when these buildings get some attention from a structural engineer, many times the engineer does not actually perform a lateral force analysis on the structure because it is assumed that certain systems are adequate without a rigorous design check. While this may be true for a majority of cases, this practice commonly leads to a lack of adequate detailing on the contract documents. As previously noted, structural materials are many times quite adequate, but the lack of attention to fastening methods and connection details have led to failures that could have easily been prevented.

While other systems can be used with shearwalls, the box system is the most commonly encountered in non-engineered and marginally engineered buildings and presents the most problems when hurricane winds are encountered. Therefore, this system will be discussed in considerable detail. The box system has horizontal diaphragms for floors and roofs which distribute loads to the shearwalls. Horizontal diaphragms are large, thin horizontal structural elements which are loaded in their plane. They receive horizontal reactions from vertical wall members such as masonry walls or stud walls which span between levels as simple beams (See Figure 6-3). Their behavior is very similar to a large horizontal beam which spans between shearwalls. Their use is not limited to use with shearwalls as a part of the box system, but this is the area where most problems are encountered.

FIGURE 6-3
TYPICAL BOX SYSTEM

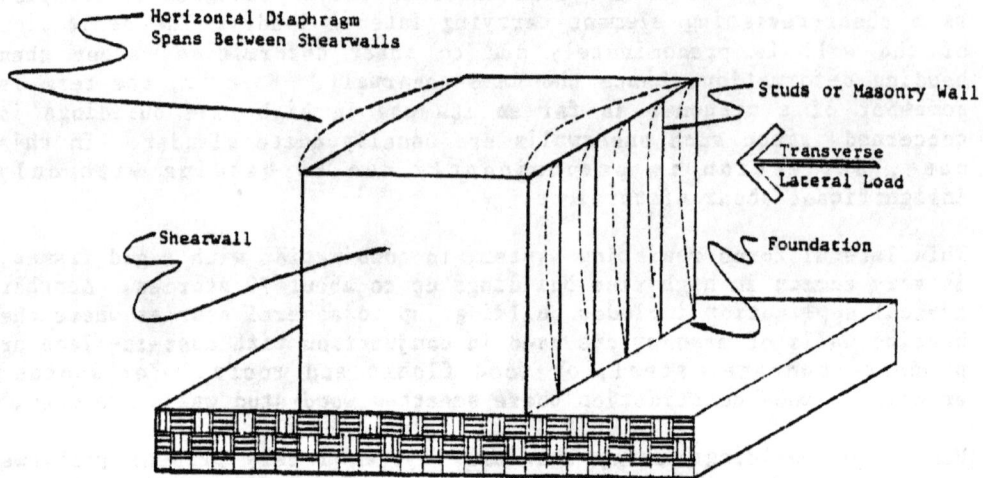

Commonly encountered diaphragms are metal decks, plywood sheathing,
particleboard sheathing, and concrete slabs. When used as a part of a
box system, these diaphragms must be adequately sized and connected to
supporting framing members and boundary elements in order to develop
their strength. One should consult various engineering standards,
texts, and literature for the structural load capacity of such
diaphragms. Most model codes contain allowable shear values for
plywood, and particleboard diaphragms used under various design
conditions. Model codes may refer to additional standards which govern
other materials as well.

Not only is strength a consideration in diaphragm design, but also
their stiffness. The stiffness of a diaphragm is related to the
amount of deflection that occurs in the diaphragm under load. There
are two general categories of diaphragm classifications as far as
stiffness is concerned: rigid and flexible. Rigid diaphragms have
very little deflection under load as compared to flexible diaphragms.
Examples of diaphragms which are commonly considered rigid are
poured-in-place concrete slabs and metal decks with poured concrete
toppings. Metal decks without concrete toppings and wood decks are
commonly considered flexible. It is important that the deflections of
diaphragms under design wind loads be kept below the point that would
cause a problem with the cracking of interior finishes and other
nonstructural elements of the building. In buildings containing
masonry walls, stiffness is important from a structural aspect as
well. Excess diaphragm deflection can cause cracks in masonry walls
that will weaken them structurally. These cracks result when the
walls are not flexible enough to move with the diaphragm as it
deflects under load. In buildings with concrete or masonry walls,
flexible diaphragms should not be used to transmit lateral forces by
rotation which occur due to the lack of coincidence between the center
of load and the center of rigidity of a building. Most model codes
explicitly prohibit this use for at least some flexible diaphragms,
such as plywood and particleboard.   Use of flexible diaphragms of
other materials to transmit lateral forces by rotation is not good
practice, even when not explicitly prohibited by codes. The principle
is the same for horizontal diaphragms of other materials and the code
can and should be enforced to exclude this application.

Calculation of diaphragm deflections is a complex matter and one should consult engineering literature. Various model codes contain maximum span-depth ratios for plywood and particleboard diaphragms which are constructed according to their tables. While these ratios help ensure that some degree of stiffness is present in such diaphragms, designers should investigate further for each given design to determine that these measures are sufficient. Engineering literature should be consulted in arriving at a conclusion. More will be said about general deflection problems later, since deflections are not limited to box structures alone.

The horizontal diaphragm can be thought of as being analogous to a horizontal beam. In wood and metal deck diaphragms, the sheathing or deck are thought of as a web of a wide flange steel beam which is considered to resist shear forces only. Chord and strut boundary members are supplied at the diaphragm boundaries. Chord boundary members are thought of as beam flanges and are assumed to develop the moment capacity of the diaphragm only. The strut boundary members are connecting members which transfer the shear reactions to the shearwalls. This principle is illustrated below in Figure 6-4. Diaphragms of precast concrete panels and poured-in-place concrete diaphragms may be considered to develop their strength in a similar fashion. Poured-in-place concrete may also be designed to use other mechanisms when governing codes do not require boundary members due to seismic considerations. Engineering judgment is advised in determining whether boundary members are needed in cast-in-place concrete members when the governing codes do not require them.

FIGURE 6-4
DIAPHRAGM LOADS

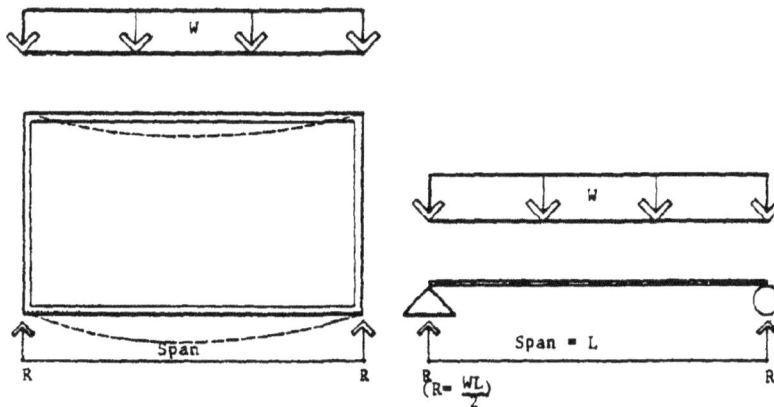

FIGURE 6-4 (Continued)
DIAPHRAGM LOADS

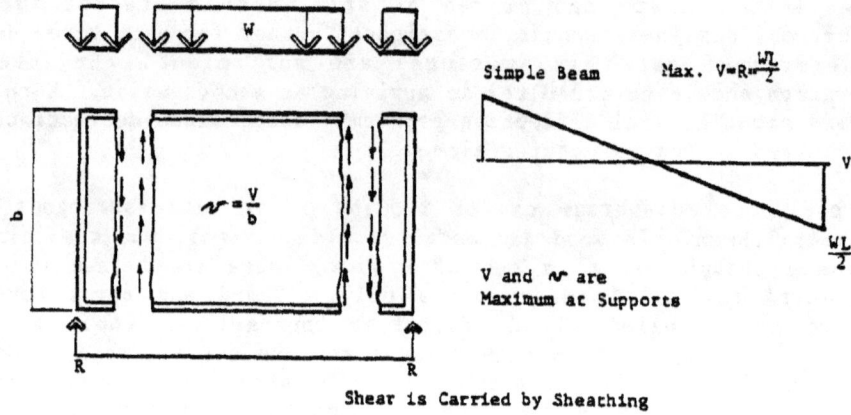

$$\mathbf{v} = \frac{V}{b}$$

Simple Beam   Max. $V = R = \frac{WL}{2}$

$\frac{WL}{2}$

V and $\mathbf{v}$ are
Maximum at Supports

Shear is Carried by Sheathing

FIGURE 6-4 (Continued)
DIAPHRAGM MOMENT

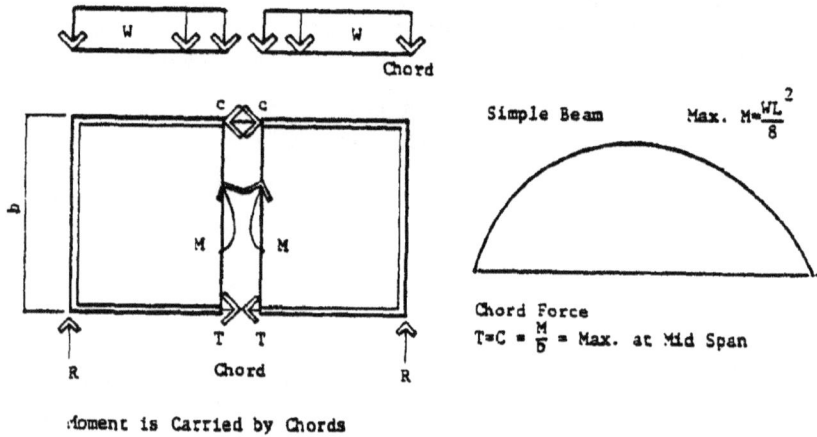

Simple Beam   Max. $M = \frac{WL^2}{8}$

Chord Force
$T = C = \frac{M}{b}$ = Max. at Mid Span

Moment is Carried by Chords

In wood frame box system construction, the top plates of the stud walls function as the chord boundary elements as well as strut elements at the ends. Hence, splices in these members as well as the sheathing connections to them are extremely important for their proper behavior under wind loads. In masonry wall construction with metal deck or wood sheathed floors and roofs, the walls themselves may serve as chords and struts by the use of steel reinforced bond beams. Shear forces from the diaphragm are usually transferred to the wall through blocking or a continuous member bolted to the wall. As noted earlier, standard engineering literature should be consulted to determine the strength of a particular diaphragm system. The basic design considerations for a horizontal diaphragm of metal deck or wood sheathing should include the following considerations:

1. Sheathing thickness
2. Layout pattern of metal deck or sheathing panels
3. Fastening methods
4. Chord design
5. Strut design
6. Diaphragm deflection
7. Tie and anchorage requirements

The importance of each of these items cannot be over emphasized, especially fastening and tie and anchorage requirements. The engineer should not only give these items top consideration in design, but should provide careful detailing on contract drawings to ensure that the required construction will be clearly presented to the contractors and inspectors. Typical details are shown below in Figure 6-5.

FIGURE 6-5
WOOD STUD WALL WITH WOOD FLOOR OR ROOF SYSTEM

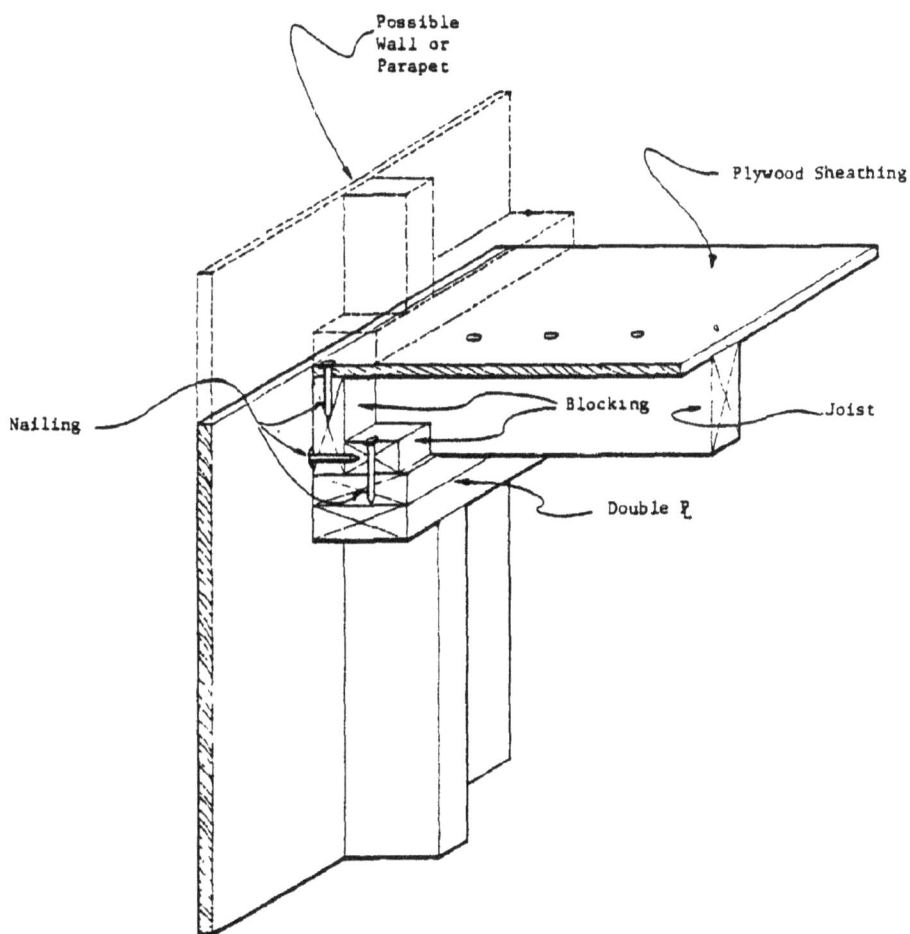

FIGURE 6-5 (Continued)
WOOD STUD WALL WITH WOOD FLOOR OR ROOF SYSTEM

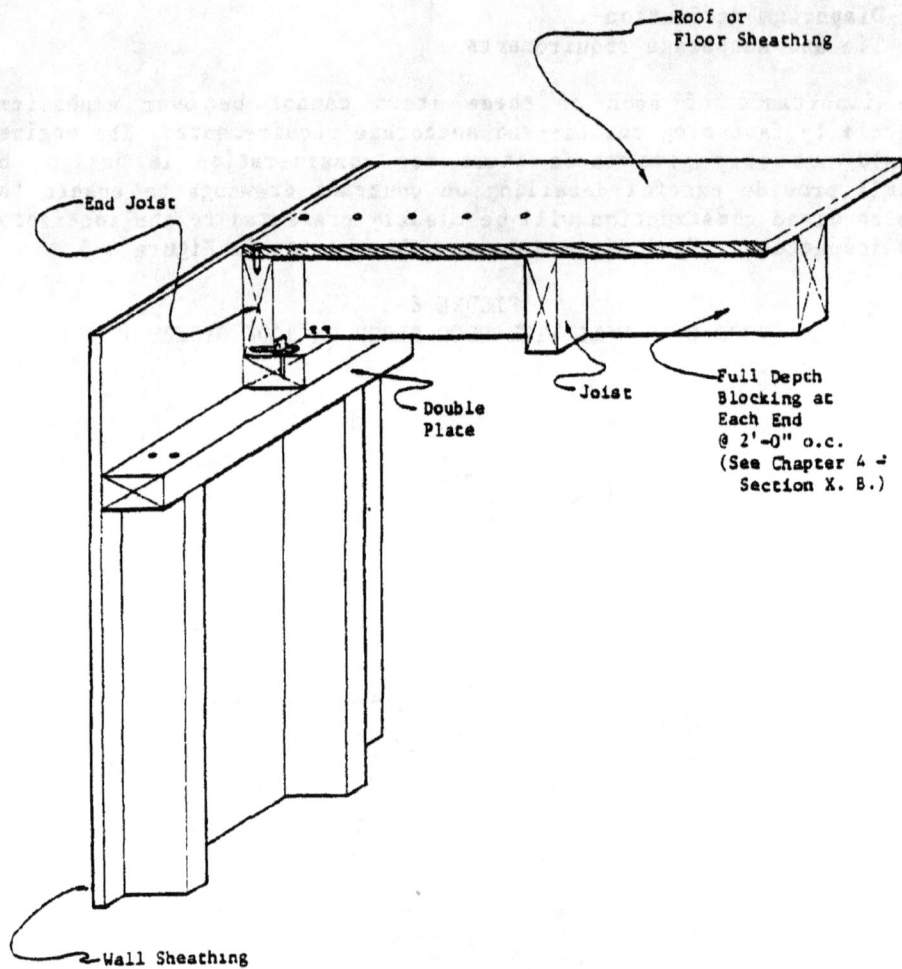

Roof or
Floor Sheathing

End Joist

Double
Plate

Joist

Full Depth
Blocking at
Each End
@ 2'-0" o.c.
(See Chapter 4 –
Section X. B.)

Wall Sheathing

FIGURE 6-5
PRECAST CONCRETE FLOOR (WITHOUT TOPPING) SUPPORTED BY MASONRY WALL

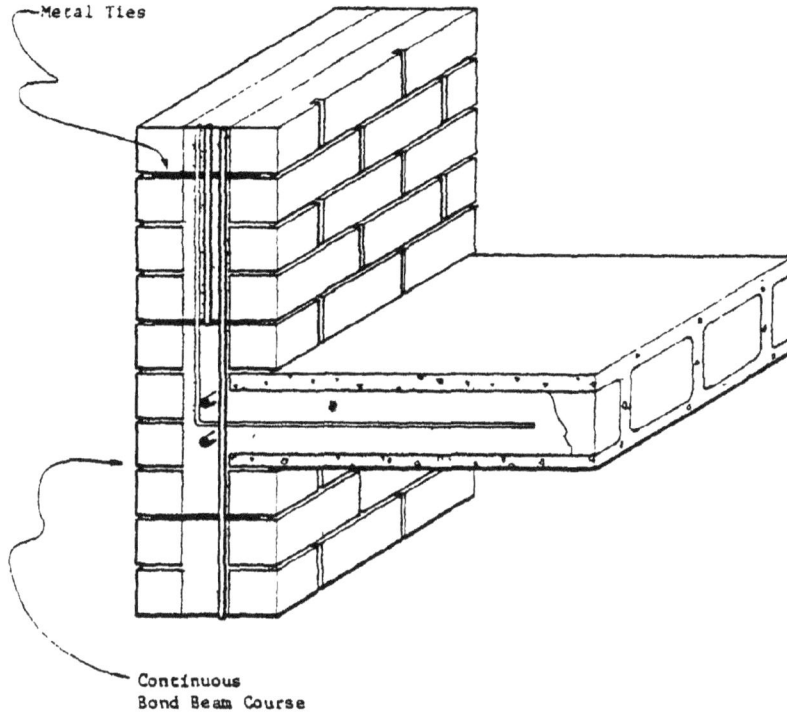

Metal Ties

Continuous
Bond Beam Course

FIGURE 6-5 (Continued)
PRECAST CONCRETE FLOOR (WITH TOPPING) SUPPORTED BY MASONRY WALL

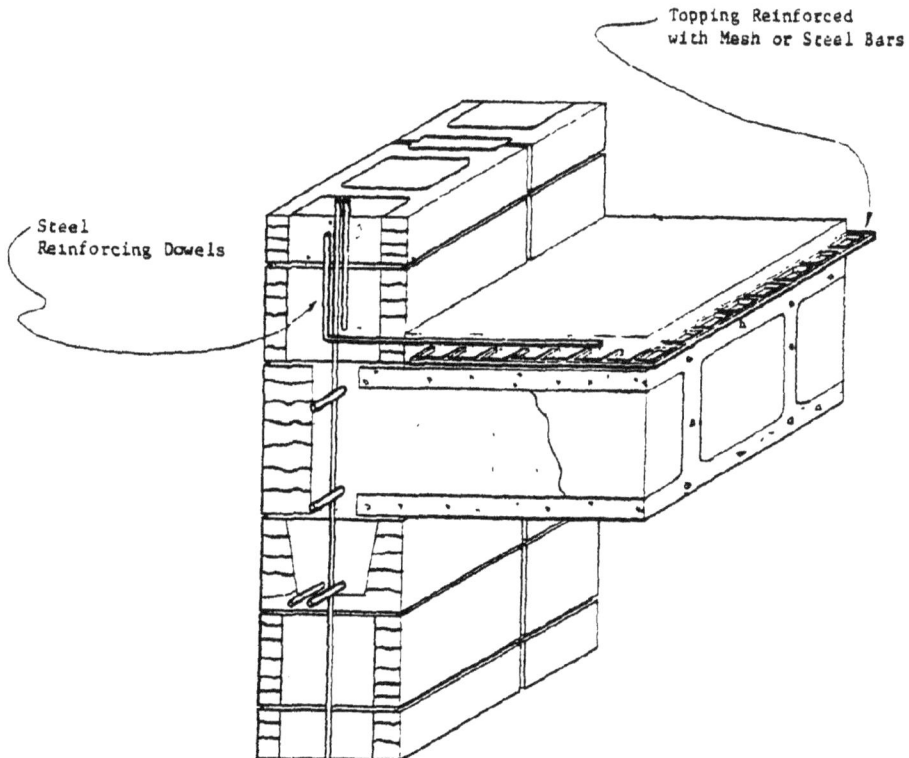

Topping Reinforced
with Mesh or Steel Bars

Steel
Reinforcing Dowels

11-6

FIGURE 6-5
MASONRY WALL WITH WOOD ROOF SYSTEM

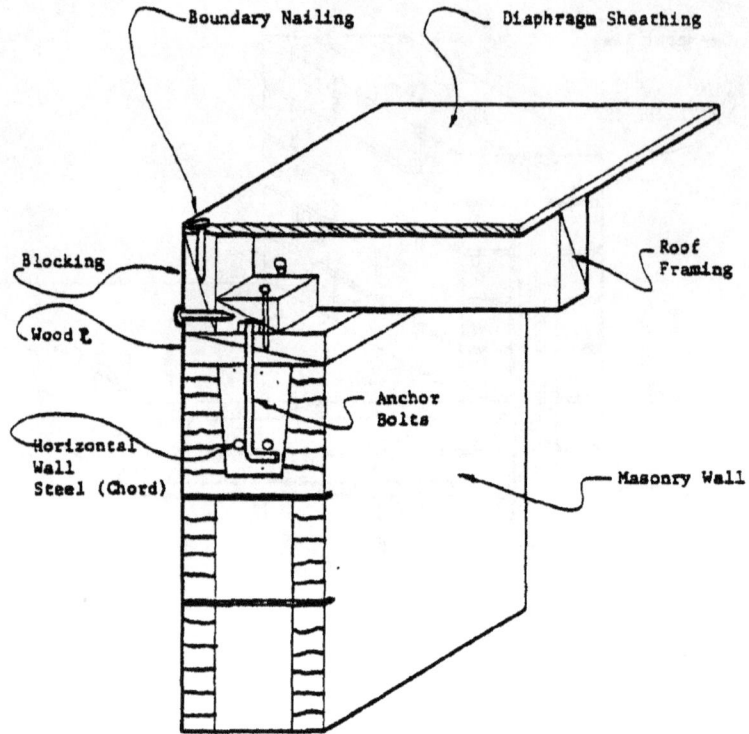

Boundary Nailing — Diaphragm Sheathing — Blocking — Wood ₤ — Roof Framing — Anchor Bolts — Horizontal Wall Steel (Chord) — Masonry Wall

FIGURE 6-5
MASONRY WALL WITH WOOD FLOOR SYSTEM

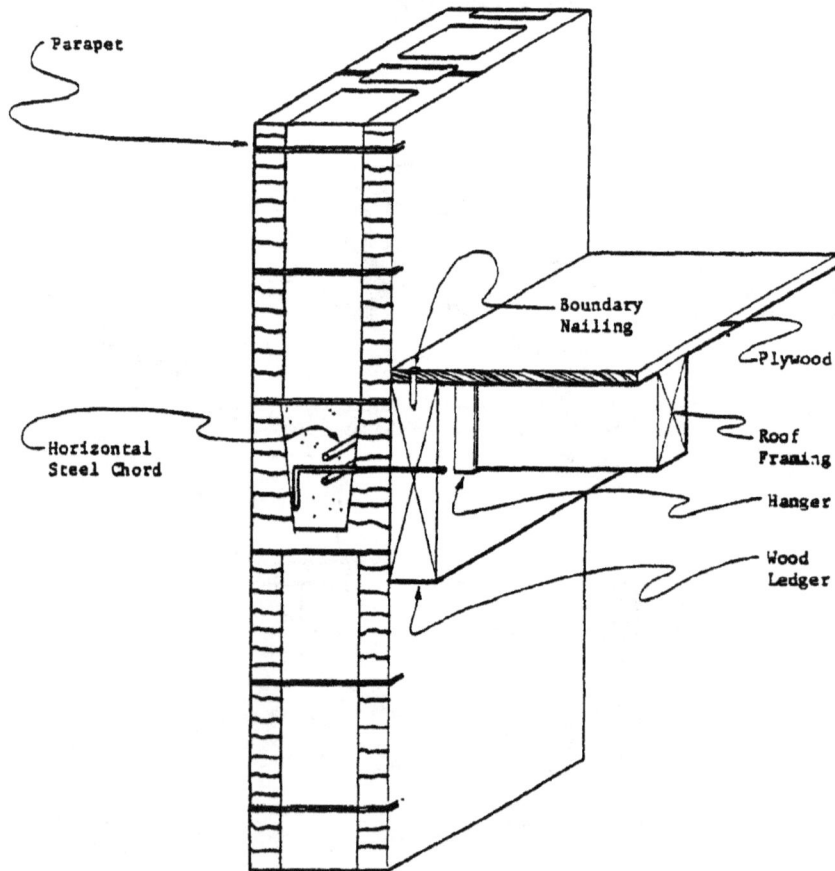

Parapet — Boundary Nailing — Plywood — Roof Framing — Hanger — Wood Ledger — Horizontal Steel Chord

Roof Covering

Ledge Angle
(Bolt to wall as
required to develop
diaphragm shear)

Open-Wed
Steel Joist

Provide required strength
at all connection levels
to transfer diaphragm shear
from deck to wall

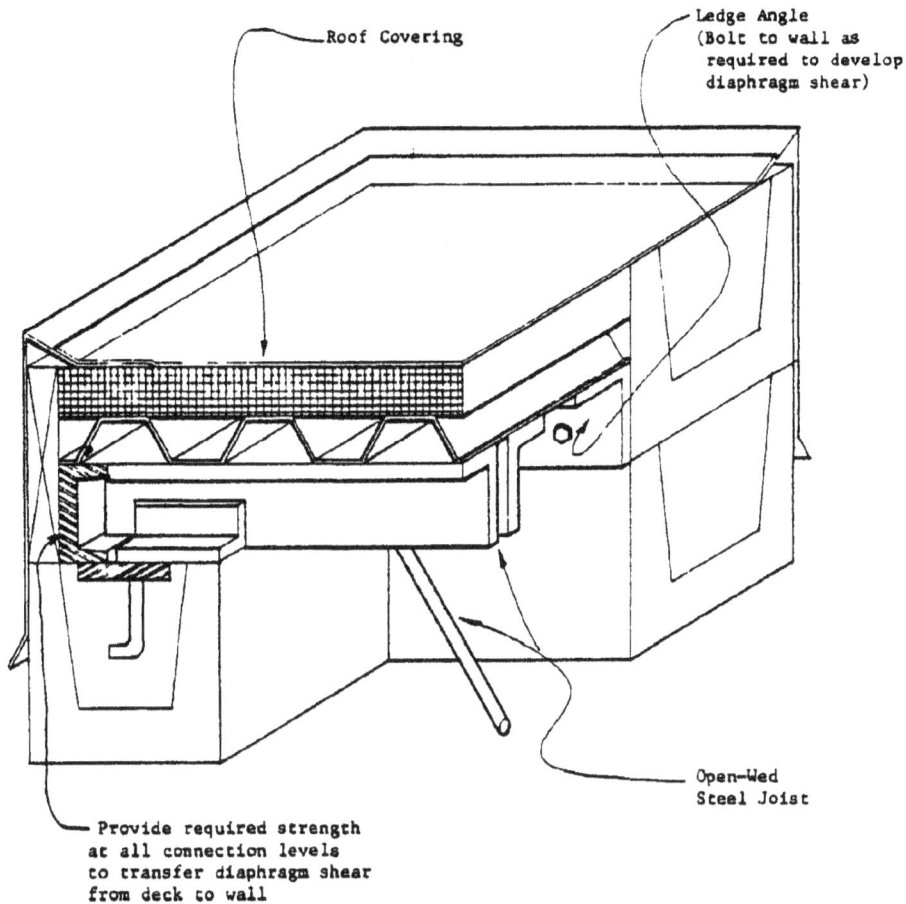

Typical problems encountered when these items are not carefully
considered may be illustrated in the failure of exterior concrete
masonry walls of buildings when subjected to hurricane force winds.
Most model codes contain maximum height-to-thickness ratios for
computing the maximum distances allowed between the lateral supports
of masonry walls. Most designers are very aware of these limitations
and comply with them by using horizontal elements such as floors and
roofs; to provide lateral support for these walls. However, many times
the diaphragm capacities of these floors and roofs are not checked to
ensure that they will actually provide the lateral support necessary.
Thus disaster results when these buildings are subjected to a
hurricane. Without a doubt, many failures which are thought to be
failures of masonry walls are actually due to failure of floor and
roof diaphragms. Under these conditions, one actually has a condition
similar to a cantilever wall from the ground level to the top of the
building since the diaphragms do not effectively brace the walls at
the floor and roof levels.

Similar failures of masonry walls may be attributed to the failure of
members supporting the diaphragm instead of the diaphragm itself.
Steel joists used in roof construction are often times not designed
for uplift loads caused by strong winds. Buckling of the bottom chord
of these joists not only means failure of these members, but also
results in a condition where the diaphragm can no longer serve its
purpose.

The remaining element of the box system is the shearwall as shown in Figure 6-6.

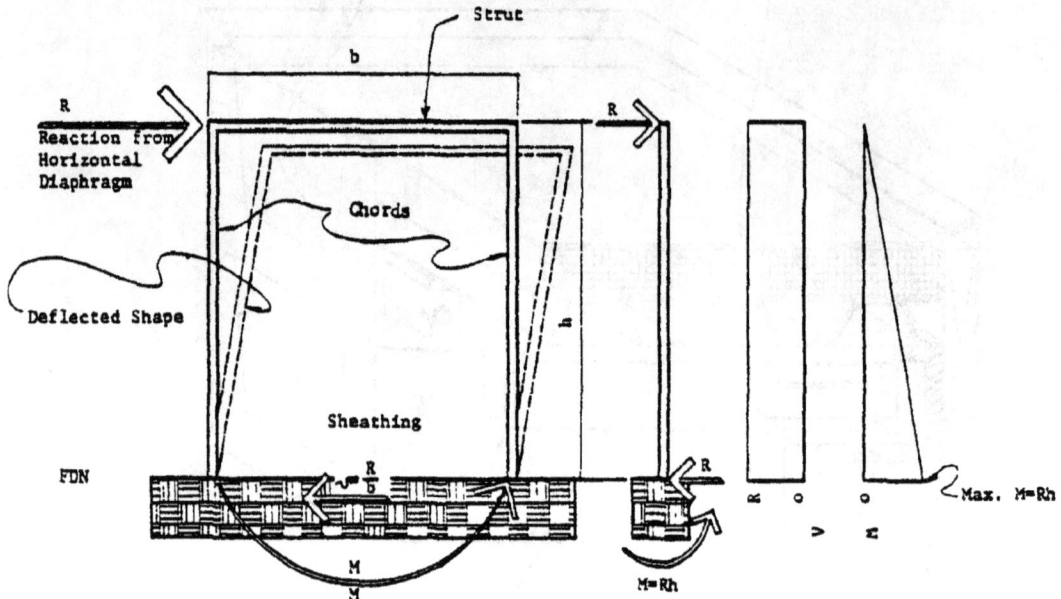

The shearwall essentially performs as a cantilever beam. Much like the diaphragm, shearwalls can be designed as boundary members such as chords resisting bending moments with interior web members resisting shear forces. This mechanism is not necessarily required by codes for masonry shearwalls or reinforced concrete shearwalls when they are not subjected to seismic forces, but it may be used by the designer. Masonry and reinforced concrete shearwalls not subject to seismic forces are very often designed as rectangular beams. As in horizontal diaphragms, engineering literature should be consulted for the load carrying capacity of these walls. Model codes contain allowable shear values for various shearwalls built of wood studs and wall membranes of materials such as plywood, particleboard, and gypsum. Model codes also refer to design specifications for other materials as well.

The basic design considerations which must be taken into account in the design of shearwalls are very similar to those of horizontal diaphragms and are as follows:

1. Sheathing type and thickness
2. Layout pattern of sheathing panels
3. Fastening methods
4. Chord design
5. Strut design
6. Shear panel proportions
7. Anchorage requirements

These general points are written around wood frame and sheathed walls.
Other materials such as masonry and concrete would be similar. One
word might be said of sheathed walls having metal studs. Information
from literature on the design of these walls is not as available as
those using wood studs and no commonly accepted design standards have
been developed. Therefore, the designer utilizing metal studs should
carefully establish the design values to be used by testing.

The importance of the seven items above is stressed again as it was in
horizontal diaphragms. Much of what was said there will not be
repeated. But, suffice it to say, the designer should be extremely
careful in making assumptions about the capacity of these walls
without actually confirming their capacity. Careful detailing on the
contract drawings is again of utmost importance to ensure that
capacity is present in the as-built structure. A brief description of
shearwall anchorage is in order however.

It is usually necessary to anchor the chord boundary members directly
to the foundation by using anchors which will transfer the tensile and
compressive forces involved. This provides anchorage of the wall
against overturning moments. Shear anchorage is usually provided by
placing anchors between the boundary members (See Figure 6-7).

FIGURE 6-7
SHEARWALL ANCHORAGE

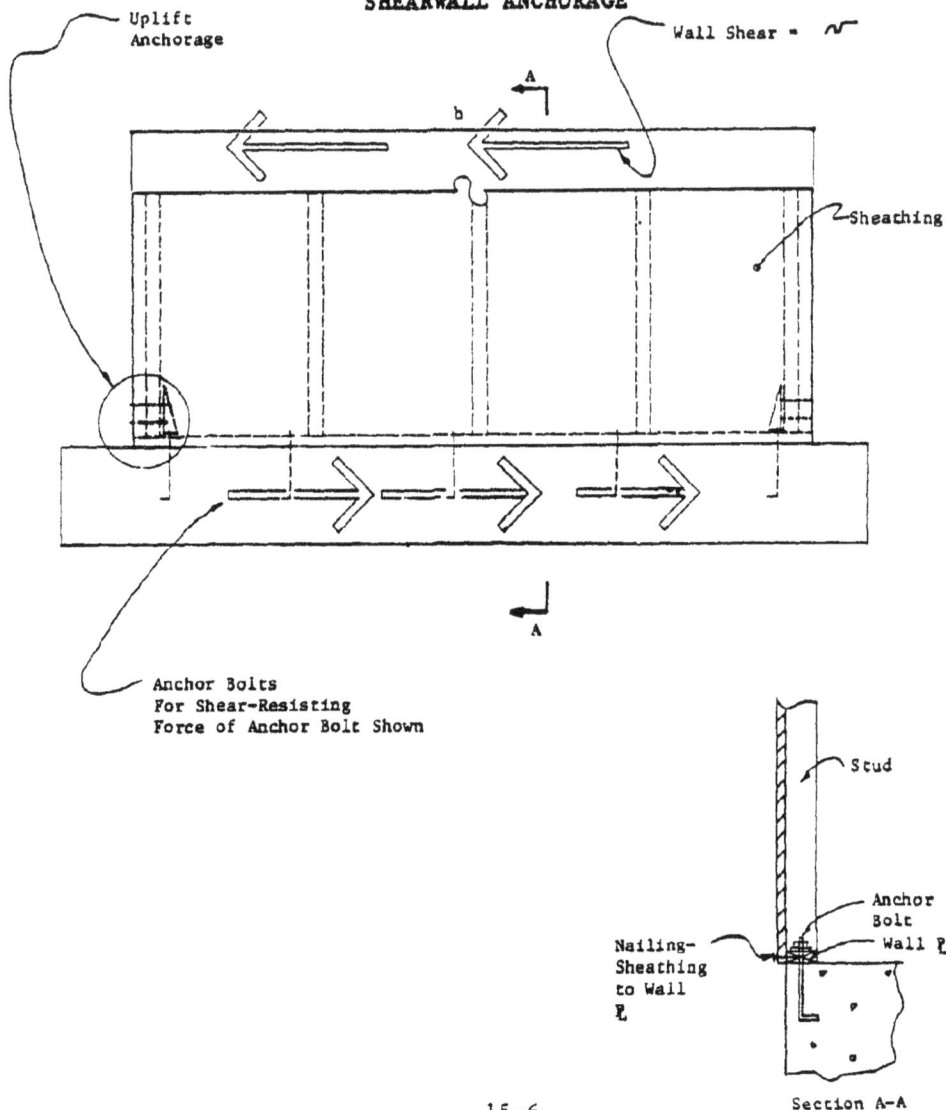

Stiffness is again a consideration in shearwalls as it was in horizontal diaphragms. Engineering literature should be consulted for methods of computing deflections and what permissible deflections might be for a given case. Model codes have maximum height-width ratios for plywood, particleboard and gypsum wallboard shearwalls which must be complied with. The designer should further check to verify that these ratios are adequate for his particular situation and provide greater stiffness when necessary.

III. CLADDING AND COMPONENTS

As noted earlier, the primary lateral force resisting systems of fully engineered buildings have performed quite well when subjected to hurricane force winds. However, performance of cladding and components has been a considerable problem.

Failures of building components at building eaves, ridges, and corners have been quite common. Current wind tunnel tests have confirmed that wind pressures in these regions are considerably larger than model codes have indicated. Not only this, these tests have also shown that the wind pressures used in the design of parts and portions should be based on the tributary area of the components involved. Components having small tributary areas should be designed to withstand higher pressures because they are more subject to localized gusts. For components having larger tributary areas, more of an averaging effect takes place and localized effects are less of a problem, thus lower design pressures are in order. Some of the model code groups have modified their wind loading requirements in very recent years to address these problems.

Failures have also been noted in buildings which are partially enclosed, open, or which have openings breached during high winds. This is primarily due to the higher internal pressures associated with these structures. Roof damage has been noted in metal buildings which have one or more sides open or where overhead doors have failed. The failure of overhead doors increases internal pressures to those of a partially enclosed building instead of an enclosed building as designed. Further failure of the roof sheeting often results. Recent wind provisions adopted by some of the model codes have sought to upgrade the codes in these areas. Few, if any, buildings which are designed by these standards have been subjected to hurricane force winds. In hurricanes as recent as Hurricane Alicia in 1983, these code provisions were very new and few buildings were being designed in accordance with them. In some codes, these updated provisions are only alternate provisions that apply to low rise buildings.

Unless openings are designed to withstand design wind forces, the building should be designed as if it were partially enclosed as well as enclosed. However, if doors and windows and their supports are designed to resist specified loads and the glass protected from wind blown missiles, the building can be considered enclosed. Glass curtain walls in high rise buildings have been broken by wind blown missiles on the windward side and the resulting internal pressure changes have caused failure of glass in the leeward curtain wall. Some designers believe that doors and other openings should also be able to resist a certain amount of leakage or they will act as openings, even when they remain in place during high winds. Further research is currently being conducted in this area.

It was noted above that wind blown missiles can be a problem, especially to glass curtain walls. Model codes currently do not address this type of load. There is considerable disagreement on whether and how they should in the future. Some possible solutions, at least for glass, are as follows:

1. Leave the codes as they are and let owners and designers determine to what extent they will design for these possibilities.

2. Eliminate the use of ballasted roofs or implement methods of controlling ballast since wind blown roof gravel has been the major cause of glass breakage in wind storms.

3. Require the use of glazing systems with laminated or other glass that is capable of withstanding debris impact.

IV. BUILDING APPURTENANCES

All building appurtenances should be designed to resist wind loads. Items such as awnings, canopies, roof structures (including mechanical equipment), satellite dishes, signs, etc. should be designed to resist wind loads and be securely anchored to the building. Failure to do so not only has led to the destruction of these items, but has created wind blown debris which has added to the hazards of other buildings. Failures of canopies over gasoline station pump islands and sign structures are frequent during high winds and bring attention to this problem. Most model building codes include these items in their wind provisions, but many times these items are not engineered for wind. Not only should the appurtenance be designed for wind loadings, the building or supporting structure must be designed for the additional wind loads which are delivered to it from such appurtenances. If such appurtenances are added to an existing building or structure, their additional loading effect should be investigated so the existing building or structure can be strengthened if necessary.

V. DRIFT

Drift is the magnitude of the displacement of the top of a building with respect to its base. This is a serviceability consideration since building movement can result in cracking of partitions and glass as well as making the building sway perceptible to the building occupants to the point of rendering the building undesirable from a user's viewpoint. These can be serious problems even though they are not associated with structural collapse. Most model codes do not have any criteria concerning building drift due to wind loads, even though some have such requirements for seismic loads. The designer should seek guidance from engineering literature in the area. Suggested maximum allowable drift is usually presented in terms of the "deflection index" - the ratio of story deflection to story height. Suggested values range from 0.0015 to 0.0035, depending on type of construction, exposure, wind requirements, etc. The designer should consult this literature and use his experience and judgment in this area. The deflection index does not address the problem of perception by the building occupants, however. Little guidance is available from literature in this area due to its many complexities.

VI. SPECIAL CONSIDERATIONS

It should be noted that the designer must be aware that the design wind loads presented in model codes apply to the majority of buildings and structures, but are not sufficient in every case. Judgment must be used when designing structures having unusual geometric shapes, dynamic response characteristics, or site locations for which channeling effects or buffeting in the wake of upwind obstructions can occur. In these cases, recognized literature must be consulted or loadings should be determined from wind-tunnel tests. When no recognized literature is available, the wind-tunnel procedure is the only option.

With the trend toward more innovative building design having to do with building shape, the use of tall, thin structures, and building at unusual sites, attention to these matters becomes more important. These new situations, and others that might occur in the future, must be approached with the awareness of the fact that a new aspect of building design has been undertaken. Its problems must be approached with a sense that problems may be present which have not been encountered previously.

BIBLIOGRAPHY

Beall, C., _Masonry Design and Detailing for Architects, Engineers, and Builders_, Prentice-Hall, Inc., Englewood Cliffs, New Jersey, 1984.

Breyer, D. E., _Design of Wood Structures_, McGraw-Hill, Inc., New York, New York, 1980.

Bowles, Joseph E., _Foundation Analysis and Design_, 3rd Edition, McGraw-Hill Book Company, New York, New York, 1982.

Building Code Requirements for Engineered Brick Masonry, Brick Institute of America, Roston, Virginia, 1969.

Building Code Requirements for Concrete Masonry Structures - ACI 531-79, American Concrete Institute, Detroit, Michigan, 1979.

Building Code Requirements for Reinforced Concrete and Commentary on Building Code Requirements for Reinforced Concrete - ACI 318-83, American Concrete Institute, Detroit, Michigan.

Council on Tall Buildings, _Monograph on Planning and Design of Tall Buildings_, 5 Vol, E. H. Gaylord and R. Mainstone, ed., American Society of Civil Engineers, New York, New York, 1980.

Design of Light Gage Steel Diaphragms, American Iron and Steel Institute, New York, New York, n.d.

Ferguson, P. M., _Reinforced Concrete Fundamentals_, 3rd Ed., John Wiley and Sons, Inc., New York, New York, 1973.

Foundation Engineering Handbook, Hans F. Winterkorn and Hsai-Yang Fang, ed., Van Nostrand Reinhold Company, New York, New York, 1975.

Gero, J. S. and Cowan, H. J., _Design of Building Frames_, Applied Sciences Publishers, LTD, London, 1976.

Handbook of Concrete Engineering, M. Fintel, ed., Van Nostrand Reinhold Company, New York, New York, 1974.

Houghton, E. L. and Carruthers, N. B., _Wind Forces on Buildings and Structures - An Introduction_, John Wiley and Sons, New York, New York, 1976.

Liu, Cheng and Evett, Jack B., _Soils and Foundations_, Prentice-Hall, Inc., Englewood Cliffs, N.J., 1981.

Luttrell, L. D., _Steel Deck Institute Diaphragm Design Manual_, Steel Deck Institute, St. Louis, Missouri, 1981.

MacDonald, A. J., _Wind Loadings on Buildings_, Applied Science Publishers, LTD., London, 1975.

Minimum Design Loads for Buildings and Other Structures - ANSI A58.1-1982, American National Standards Institute, Inc., New York, New York, 1982.

National Design Specification for Wood Construction, National Forest Products Association, Washington, D. C., 1982.

Nonreinforced Concrete Masonry Design Tables, National Concrete Masonry Association, Arlington, Virginia, 1971.

Norris, C. H., Wilbur, J. B., and Utku, S., *Elementary Structural Analysis*, 3rd Ed., McGraw-Hill, Inc., New York, New York, 1976.

Notes on ACI 318-83, Building Code Requirements for Reinforced Concrete with Design Applications, G. B. Neiville, ed., Portland Cement Association, Skokie, Illinois, 1980.

Salmon, C. G. and Johnson, J. E., *Steel Structures - Design and Behavior*, 2nd Ed., Harper and Row, Publishers, New York, New York, 1980.

Simplified Design of Reinforced Concrete Buildings of Moderate Size and Height, G. B., Neville, ed., Portland Cement Association, Skokie, Illinois, 1984.

Specification for the Design of Cold-Formed Steel Structural Members, American Iron and Steel Institute, New York, New York, 1980.

Specification for the Design and Construction of Load-Bearing Concrete Masonry, National Concrete Masonry Association, Herndon, Virginia, 1970.

Specification for the Design, Fabrication and Erection of Structural Steel for Buildings, American Institute of Steel Construction, Inc., Chicago, Illinois, 1978.

Standard Building Code, 1985 Ed., Southern Building Code Congress International, Inc., Birmingham, Alabama, 1985.

Standard Specifications, Load Tables and Weight Tables for Steel Joists and Joist Griders, Steel Joist Institute, Myrtle Beach, South Carolina, 1984.

Structural Engineering Handbook, E. H. Gaylor and C. N. Gaylord, ed., 2nd Ed., McGraw-Hill Book Company, New York, New York, 1979.

The BOCA Basic/National Building Code/1984, 9th Ed., Building Officials and Code Administators International, Inc., Country Club Hills, Illinois, 1984.

Uniform Building Code, 1985 Ed., International Conference of Building Officials, Whittier, California, 1985.

Vanderbilt, M. D., *Matrix Structural Analysis*, Quantum Publishers, Inc., New York, New York, 1974.